THE GREAT
BEYOND

*Higher Dimensions, Parallel Universes,
and the Extraordinary Search for a
Theory of Everything*

PAUL HALPERN

WILEY

John Wiley & Sons, Inc.

Published by John Wiley & Sons, Inc., Hoboken, New Jersey
Published simultaneously in Canada

Cartoon p. 8, courtesy of the Library of Congress. Figures pp. 19, 22, 24, 37, 48, 52, 75, 127, 253, 265, 273 concept by Paul Halpern. Rendered by Jerry Antner, RichArt Graphics; p. 56, concept by Paul Halpern, based on an illustration by Edwin Abbott, rendered by Jerry Antner, RichArt Graphics. Photos pp. 27, 93, 134, 168, 187, 194, 242 © Paul Halpern; pp. 29, 42, 143 courtesy of the Archives, California Institute of Technology; p. 154, by John Hagemeyer, courtesy of the Archives, California Institute of Technology; p. 106, courtesy of the University of Göttingen; pp. 117, 122, 197, courtesy of the Niels Bohr Archive; p. 163, courtesy of Syracuse University; p. 233, courtesy of Claud Lovelace and Rutgers University; pp. 237, 254 courtesy of the American Institute of Physics, Emilio Ségre Visual Archives, Physics Today Collection

For general information about our other products and services, please contact our Customer Care Department within the United States at (800) 762-2974, outside the United States at (317) 572-3993 or fax (317) 572-4002.

Wiley also publishes its books in a variety of electronic formats. Some content that appears in print may not be available in electronic books. For more information about Wiley products, visit our web site at www.wiley.com.

Library of Congress Cataloging-in-Publication Data:

Halpern, Paul, date.
    The great beyond : higher dimensions, parallel universes, and the extraordinary search for a theory of everything / by Paul Halpern.
        p. cm.
    Includes bibliographical regerences.
    ISBN 0-471-46595-X (cloth : alk. paper)
        1. Physics—Philosophy.   2. Relativity (Physics)   3. Particles (Nuclear physics)
4. Cosmology.   I. Title.
QC6.H273   2004
530'.01—dc22   2004004687

Printed in the United States of America

10 9 8 7 6 5 4 3 2 1

Dedicated to Michael Erlich and Frederick Schuepfer,
for their great friendship and inspiration over the years.

Non-Euclidean calculus and quantum physics are enough to stretch any brain; and when one mixes them with folklore, and tries to trace a strange background of multi-dimensional reality behind the ghoulish hints of the Gothic tales and the wild-whispers of the chimney corner, one can hardly expect to be wholly free from mental tension.

—H. P. LOVECRAFT, "The Dreams in the Witch-House" (1933)

Contents

Acknowledgments   vii

Introduction: The Kaluza-Klein Miracle   1

1   The Power of Geometry   10

2   Visions of Hyperspace   26

3   The Physicist's Stone: Uniting Electricity, Magnetism,
    and Light   61

4   Getting Gravity in Shape   84

5   Striking the Fifth Chord: Kaluza's Remarkable
    Discovery   101

6   Klein's Quantum Odyssey   114

7   Einstein's Dilemma   139

  8   Truth under Exile: Theorizing at Princeton   158

  9   Brave New World: Seeking Unity in an Age
      of Conflict   179

 10   Gauging the Weak and the Strong   206

 11   Hyperspace Packages Tied Up in Strings   231

 12   Brane Worlds and Parallel Universes   267

 Conclusion: Extra-dimensional Perception   290

 Notes   299
 Further Reading   314
 Index   319

Acknowledgments

I wish to acknowledge the generous support of the John Simon Guggenheim Memorial Foundation, which has provided me with the time and resources for this research. The fine examples set by past and current Guggenheim fellows have offered a constant source of inspiration for my work.

Several notable historians of science offered me valuable advice for this project. I thank them sincerely for their help. These include Gerald Holton, John Stachel, Martin J. Klein, and Daniela Wünsch. Thanks also to Jürgen Renn and Anne J. Kox, who offered their friendly support. The Philadelphia Area Seminar on the History of Mathematics has been a great source of guidance. I thank the members of the group for their comments, and acknowledge the critical assistance of David Zitarelli, Robert Jantzen, Paul Pasles, Jim Beichler, and Thomas Bartlow. I would also like to honor the memory of my advisor, Max Dresden, who brought the history of science to life in his colorful lectures.

Getting to meet and interview the legendary physicist John Wheeler was a special thrill. I thank him for his insights and hospitality, and thank Kenneth Ford for arranging the interview, as well for as his own comments. I was also fortunate to have valuable discussions with many other leading physicists, whom I thank sincerely for their time, including Bryce DeWitt, Gary Gibbons, Paul Steinhardt, Raman Sundrum, Savas Dimopoulos, Alan Chodos, Lisa Randall, Claud Lovelace, and Paul Wesson. Thanks to Peter van Nieuwenhuizen for useful references. Jeff Harvey kindly gave me permission to quote from his song parody. For the use of Einstein's quotations, I recognize the permission granted by the Einstein Papers Project, California Institute of Technology, and the Albert Einstein Archives at the Hebrew University of Jerusalem in Israel.

Many thanks to Stanley Deser, who shared his insights and took the time to offer me a window into the world of Oskar Klein. A special word of appreciation to Bernard Julia and Eugene Cremmer of the École Normale Supérieure in Paris for their hospitality and advice during my visit. I would also like to thank astronomers Harry Shipman and Steven Dick, as well as physicists Dieter Forster and Justin Vasquez-Poritz for their suggestions and support. Mathematicians Samuel J. Patterson and Martin Kneser of the University of Göttingen were extraordinarily kind in describing the history of their department, especially during the time of Kaluza. Psychiatrist John Smythies was of great help for his comments and references about dimensionality and the mind. For elucidating the background of his multimedia pieces and their connections to higher dimensions, I am grateful to filmmaker Peter Rose.

The Bergmann family has been of enormous help to my research. It was a distinct pleasure having the opportunity to speak to Peter Bergmann in summer 2002. With his passing, the physics community has lost an exceptional scholar, organizer, and educator. I very much appreciate the kindness of Ernest Bergmann in answering many questions about his father. I also thank John Bergmann, Joshua Goldberg, and Linda Jean Owens (Peter Bergmann's caretaker when he was ill) for their help. Engelbert Schucking, one of Peter Bergmann's closest friends and a great scholar of relativity, was kind enough to offer his valuable insights.

I would like to thank Robert S. Cox, Valerie-Anne Lutz, and the staff of the Manuscripts Library at the American Philosophical Society for their friendly assistance. The staff of the Department of Rare Books and Special Collections at Princeton University, including Margaret Rich and AnnaLee Pauls, have been of tremendous help in making it possible for me to examine the Albert Einstein Duplicate Archives. The libraries of the University of Pennsylvania, Temple University, the University of Göttingen, Cambridge University, and the London Metropolitan Archives have also been of great use in my research. Thanks also to David Rose, the librarian at the City of London school, for his generous assistance, for showing me the Edwin Abbott collection and for taking me on a tour of the school. David Martin, head of mathematics for the school, was also very helpful.

I would like to express my appreciation to Felicity Pors, Finn Aaserud, and the staff of the Niels Bohr Archive for permitting me to examine the Oskar Klein papers, and for their great hospitality during my visit to Copenhagen. I also appreciate the help of Harry Leechburch, who offered kind assistance when I viewed the Paul Ehrenfest material at the Museum Boerhaeve in Leiden.

Many thanks to the leadership of the University of the Sciences in Philadelphia, including Philip Gerbino, Barbara Byrne, C. Reynold Verret, Charles Gibley, Margaret Kasschau, Elizabeth Bressi-Stoppe, Allen Misher, Anthony McCague, John Martino, Joseph Trainor, Robert Boughner, and Stanley Zietz, for making it possible for me to have a productive sabbatical year, free of the distractions of committee work and teaching. Thanks also to faculty members Roy Robson, David Traxel, Barbara Bendl, David Kerrick, Salar Alsardary, Durai Sabapathi, Tarlok Aurora, Ping Cunliffe, Amy Kimchuk, Lia Vas, Greg Manco, Bernard Brunner, Charles Samuels, Roy Schriftman, Anatoly Kurkovsky, Robert Field, Ara der Marderosian, and others. I very much appreciate the help of Judeth Kuchinsky, the masterful administrative assistant of my department, and of Shannon Stoner, interlibrary loan librarian extraordinaire.

Thanks to Arlene Renee Finston for her critical photographic assistance and to Jerry Antner of RichArt Graphics for his superb renderings of the illustrations.

My wonderful agent, Giles Anderson, has been a particular source of support and inspiration. Thanks to my editor, Eric Nelson, and the staff of John Wiley & Sons, for their useful advice and assistance.

Finally, I'd like to express my great appreciation to my family and friends during a time of intense effort. My parents, Stanley and Bernice; my in-laws, Joe and Arlene; Herb and Peggy, as well as Richard, Anita, Alan, Beth, Kenneth, and other family members, have all been very supportive. Elana and Roy Lubit, Scott Veggeberg, Marcie Glicksman, Dorothee and Helmut Funke, Simone Zelitch, Doug Buchholz, Lindsey Poole, Greg Smith, Donald Busky, Fred Schuepfer, Dubravko Klabucar, Mitch and Wendy Kaltz, Fran Sugarman, and Michael Erlich have provided lots of friendly help. Above all, thanks to my wife, Felicia, and my sons, Eli and Aden, for all their love throughout the years.

Introduction

The Kaluza-Klein Miracle

The formal unity of your theory is astonishing.

—ALBERT EINSTEIN, letter to Theodor Kaluza, 1919

It is an elegant idea, fashioned in the magnificent lathe of mathematical insight. A radical idea—one that has inspired revelry as well as derision. A persistent idea, durable enough to have outlasted two world wars, as well as the twists and turns of twentieth-century science. And it is a compelling idea: the intriguing notion that nature looks most complete when wrapped up in a garb of extra dimensions.

Though it has at various times been triumphed, mocked, misinterpreted, and ignored, the concept of higher dimensions beyond space and time has become a central feature of modern theoretical discussion. If it is true, it would mean that the world we perceive is only a fraction of a greater invisible reality. Length, width, breadth, and duration would be supplemented by unseen directions, outside the range of our senses.

The scientific community is traditionally a cautious lot. It is resistant to change, unless the arguments cut deep. Attractive mathematical notions, considered in the abstract, do not wield enough of an axe to sever long-held conceptions. To postulate realms beyond the scope of the familiar requires firm physical justification. Theorists' current interest in extra dimensions has emerged from a sense that taking such a bold step is the best (and perhaps the only) way of unifying all of the forces of nature into a single, cohesive expression.

Science has revealed four fundamental natural forces. The best understood of these is electromagnetism. In the nineteenth century, physicist James Clerk Maxwell successfully modeled its behavior through a simple set of equations. Its properties on the smallest scale were fully explored in the mid-twentieth century through the theory of quantum electrodynamics (QED). One of the most successful theories in the history of science, QED offers the ability to understand a full range of electromagnetic processes, from the collision profiles of hot, charged particles to the magnetic properties of ultra-cool superconductors.

Gravitation, another fundamental force, was famously described by Isaac Newton in the seventeenth century. Then, in the twentieth century, Einstein's general theory of relativity reinterpreted gravity in terms of the geometry of space and time. Unlike electromagnetism, however, gravitation is little understood on the tiniest scales. No successful theory of gravity incorporates the quantum principles known to guide atomic and subatomic behaviors. The development of a quantum theory of gravity, as successful in its predictive powers as QED, remains an outstanding goal of physics.

The weak interaction was first discerned through the process of radioactive decay, discovered in 1896 by Henri Becquerel. However, it wasn't identified as a separate force until the mid-twentieth century. Gradually physicists realized that a particular type of nuclear transformation called beta decay, involving the disintegration of neutrons into protons, electrons, and neutrinos, could be mitigated only by means of a new force. This was named the weak interaction (also known as the weak nuclear force) to contrast it with another, more powerful force, called the strong interaction, proposed around the same time.

The strong interaction (also known as the strong nuclear force) emerged as an explanation of why atomic nuclei do not fall apart. Something, physicists realized, must bind the protons and neutrons that form the centers of atoms together. Furthermore, that force must be powerful enough to counteract the force of electrical repulsion that protons feel when they are packed closely together. Like charges repel, unless something else keeps them together. That something is the strong force.

Because the strong force acts on subatomic scales, researchers realized that its effects could only be understood through quantum theory. In the past few decades, scientists have advanced a micro-scopic description of the strong force, called quantum chromody-namics (QCD), that has met with some success. Calculations in QCD are much more difficult than in QED, limiting scientists' ability to test all of the ramifications of the theory.

Like stubborn-minded brothers, each of the four natural forces behaves in its own characteristic way. Gravity, as a distortion of the space in which particles move, affects all types of matter. Electro-magnetism, in contrast, concerns itself only with charged objects, such as positively charged protons or negatively charged electrons. A neutral body, such as a neutrino or a neutron, can lie like a ghost on a busy highway, directly in the path of an immense electromagnetic field, and never notice anything at all. The strong force is even more finicky. It excludes an entire class of particles, called leptons, and embraces only another group, called hadrons. Leptons include elec-trons, neutrinos, and more massive particles called muons and tauons. Hadrons are particles, such as protons, neutrons, and many others, that are built up of even smaller constituents, called quarks. Finally, the weak interaction participates in only a limited group of particle transformations—beta decay, for instance.

Moreover, the four forces act with different strengths over dis-tinct ranges of action. Electromagnetism and gravity are effective over very large distances, dropping off in strength at a relatively slow rate compared to the other two forces. That is why a compass can detect Earth's magnetic poles, thousands of miles away, and why oceans can respond to the gravitational pull of the Moon. The weak and strong interactions, on the other hand, act only over very lim-ited ranges—namely at most on the scale of an atomic nucleus. Two protons, located a quarter inch apart, would feel no measurable strong attraction, only an overwhelming electrical repulsion. Some-thing would have to move them, against their "will," many trillion times closer for them to experience the pull of the strong force. That is why the binding together of protons and neutrons in the process of nuclear fusion takes place only under high-pressure conditions, such as in the heart of the Sun. The immense pressure pushes the particles toward each other, enabling them to feel the strong inter-action and stick together.

Once two particles are close enough for the strong interaction's glue to bind them, they stick together with incredible strength. The strong interaction, at short range, is the mightiest of the forces. It flexes far more muscle than electromagnetism. Within the tight confines of the nucleus, electromagnetism gets sand kicked in its face. That is why the nuclei of the most common atoms, from hydrogen to iron, are so stable.

The least powerful force, by far, is gravity. In a competition at a cosmic gym, it wouldn't even have the strength to write down its score. Even the weak force, humbled by the strong and electromagnetic forces, could trample gravity with one meager breath. Gravity is more than a quadrillion quadrillion times weaker than any of the other forces. One indication of the strength of electromagnetism compared to gravity, for instance, is that an average-sized bar magnet can easily overpower the whole of Earth's gravitational attraction and lift a set of paper clips off the ground. The dilemma of why gravity is so much weaker than the other interactions is called the hierarchy problem.

The four forces certainly make odd brothers. Yet most physicists firmly believe that they share common parentage. At the time of their birth, in the fiery first instants of the universe, they all looked the same. Each force had the same range, strength, and ability to interact with particles. Somehow, though, in the changing environment that marked the passage of time, each force went its separate way and acquired its own characteristics. As the universe cooled, these distinct properties froze into place, like the varied shapes of ice crystals forming on a frigid window. Science has already proven the fraternity of two of the four natural forces. In the 1960s, the physicists Steven Weinberg and Abdus Salam (extending the work of Sheldon Glashow) developed a unified explanation for the electromagnetic and weak interactions, known as the electroweak model.

Theorists would love to develop an all-encompassing "Theory of Everything" that includes all four natural forces. Yet gravity is so different from the others that physicists have been forced to reach beyond the bounds of observability in efforts to include it. It is possible that the strong and electroweak interactions could live in a house similar to the one Weinberg and Salam built. But gravity residing with its brethren seems to require a colossal new extension, and that extension is a universe with higher dimensions.

The notion of unification through hyperspace (more than three spatial dimensions) has been called the Kaluza-Klein miracle. The name stems from two of its original proposers in the 1910s and 1920s: German mathematician Theodor Kaluza and Swedish physicist Oskar Klein. As originally constructed, their theories applied only to the unification of electromagnetism with gravity; the other two forces were yet unknown. In recent decades, such models, extended in attempts to include the remaining interactions, have been the subject of countless research articles and talks. In the past few years alone, theories involving variations of the Kaluza-Klein approach have received more scientific citations than virtually any other subject in theoretical physics. As prominent theorist Gary Gibbons wrote, "No one who has worked through the mathematics of Kaluza and Klein's construction can ever forget its haunting beauty, and despite its experimental limitations . . . the basic idea has come to dominate all current attempts at unifying the gravitational with the electromagnetic, weak and strong interactions."[1]

Theodor Kaluza came upon the idea of higher dimensional unity while working in the lowly position of privatdozent (unpaid lecturer) at the University of Königsberg. In Germany at the time, in terms of status, privatdozents were to professors as stunt doubles today are to leading Hollywood actors. Privatdozents drew their only income from collecting part of the fees that students paid for each lecture, as well as assisting the professors in other ways around the university. Unlike professors, they had no office or prestige. If a privatdozent was unlucky enough or unpopular enough that few students attended his class, and if he couldn't supplement his income otherwise, he could literally starve. Under these difficult conditions, nevertheless, Kaluza maintained remarkable creativity.

One day, Kaluza was sitting in his study at home, considering a variation of Einstein's then-new general theory of relativity. In an arrangement that comforted him, his nine-year-old son sat in another part of the room and watched him while he worked. Kaluza thought it would be interesting to rewrite Einstein's equations in a hypothetical five-dimensional universe, perhaps in the same way that Ravel found it engaging to rework Mussorgsky's "Pictures at an Exhibition," with a full orchestra instead of just a piano. What new features would the additional element reveal? Kaluza wondered.

Suddenly, Kaluza had a revelation. By adding the extra dimension, he found that his reconstituted equations contained not only Einstein's theory of gravitation, but also Maxwell's theory of electromagnetism. To his son's surprise, he froze momentarily in place, then stood up abruptly and—in his own eureka shout—began to hum a Mozart aria.

Kaluza was clearly excited. He believed he had found the key to unifying all of nature. He submitted his findings to Einstein, then editor of a prestigious German journal. Einstein was initially delighted by the idea, then had some misgivings. He held up its publication for two years while he sent Kaluza a number of suggestions. Finally, deeming the notion too important for it never to appear, he published it in 1921.

Three years later, Oskar Klein, who was unfamiliar with Kaluza's paper, independently discovered the same idea. At the time, Klein was working at the University of Michigan, teaching courses in basic physics. As part of a research project, he was examining the motion of particles in fields. A field provides a map of the amount and direction of force per particle at given points throughout space. It indicates where certain forces act stronger or weaker on particles. Klein studied how charged particles moved under the simultaneous influence of electromagnetic and gravitational fields. While performing this investigation, he realized that he could encompass both electromagnetism and gravitation with a single set of five-dimensional equations.

Klein's son-in-law, physicist Stanley Deser, likes to joke about the reason Klein came to this conclusion: "I always say that he invented Kaluza-Klein theory in order to lower his teaching load. . . . He didn't want to teach both electromagnetism and gravity the following semester so he created Kaluza-Klein theory."[2]

The reason Klein usually shares credit, even though his was a "rediscovery," is that he put his own stamp on the concept. After Klein returned to Europe, he showed physicist Wolfgang Pauli his work. Pauli informed him about Kaluza's paper. Klein was devastated at first. After recovered from the shock, he decided to publish his work anyway, emphasizing a novel interpretation of the fifth dimension within the context of quantum theory. In Klein's version, the fifth dimension is wrapped in a tight loop, like a thread around a tiny spool. Its length depends on several constants of nature, including the charge of the electron, the speed of light, the gravitational constant, and Planck's constant (that sets the scale for quantum phenomena). Calculating the value of this length, Klein found it to be an exceedingly small number, beyond the limits of measurability. In this manner, he found a way of physically describing the fifth dimension and justifying the fact that it has never been detected. Hence, higher-dimensional unification of forces has come to be known as Kaluza-Klein theory.

Considering its importance, one would think that the proposers of the theory would have spent their careers trumpeting it. Not so. Ironically, after his one short paper, Kaluza published no other writings on the subject. Klein wrote several papers, then had a change of heart in which he, along with Pauli, drank to the death of his own idea. Advances in quantum theory had convinced him, for the time being, that the fifth dimension was no longer necessary. Much later, Klein found renewed interest in five-dimensional theory, returning to the subject several other times in his career with a variety of novel interpretations. However, he was much better known, at least in his lifetime, for his other contributions to physics. Sadly, Klein died shortly before his theory experienced a great revival in the late 1970s and 1980s.

In recent decades, historians of science have come to realize that, as in the discovery of the New World, there was a "Leif Eriksson" who crossed the waters before the more famous voyages. Kaluza and Klein, as it turns out, weren't exactly the first to set foot on the shores of higher dimensions. Gunnar Nordström, a Finnish physicist, had planted his flag there several years before them. His contribution was virtually lost to history before it was excavated in the 1980s. Why isn't it called the "Nordström miracle?" Perhaps for the same reason that Columbus got all the press. While Nordström's achievement was a lone and tenuous settlement, built on shaky ground (he based it on a flawed theory of gravity), the theory put forward by Kaluza and then honed by Klein inspired many others to follow. These included prominent physicists such as Einstein. Einstein spent the latter part of his career investigating various unified field theories, including five-dimensional approaches.

Yet even the great Einstein went back and forth over whether he believed in extra dimensions, and, if they existed, what they could possibly look like. Contrary to popular myth, Einstein was not the resolute thinker who created ironclad theories every time he put his thoughts to paper. His working career was full of many aborted attempts, astonishingly sudden changes of heart, and curious episodes when he would say one thing, then do something completely different, as scientific historians Abraham Pais and John Stachel have aptly pointed out. Einstein's genius lay in his unique perspective and stubborn persistence as much as in the quality of his published writings, especially in his later years.

Einstein approached the idea of the fifth dimension like a dieter with a sweet tooth. At first, when Kaluza sent him his original paper, Einstein found it luscious and enticing. Then, realizing the extra metaphysical poundage it would add to general relativity, he politely resisted for a time. Throughout the 1920s, he nibbled a bit on some of the theory's premises, but refused to devour its conclusions. Instead, he mainly feasted on other unified approaches—ones that

In this 1930 cartoon by Clifford Berryman, Einstein tells Congress,
"I'll stick to relativity and the fourth dimension!"

contained space, time, and nothing beyond. In the early 1930s, along with his assistant Walther Mayer, he developed a kind of dietetic version of Kaluza-Klein, one that made use of its tasty benefits without explicitly adding the weight of extra dimensions. Only in the late 1930s did Einstein become a Kaluza-Klein gourmet, fully savoring its delicious concoction. Until, that is, the 1940s, when he finally abandoned it like an overstuffed diner.

Einstein's discomfort with the idea of unseen extra dimensions is quite understandable, given his predilection toward tangible, testable descriptions of nature. He shared with most of his scientific contemporaries a disdain for the occult, and had to grapple with the fact that much of the public in his day associated higher dimensions with the world of the spirit. After all the progress made by the tried and true scientific method, postulating dimensions that couldn't directly be measured seemed a step backward. Nevertheless, at various times in his career, he was willing to set aside his aversion to the notion of imperceptible realms in hopes of fulfilling his dream of unification.

This is the story of Kaluza, Klein, Einstein, and many others as they grappled with the promising but unsettling implications of establishing nature's unity through higher dimensions. It is a chronicle that began in the first decades of the twentieth century, in an age in which quantum physics and relativity were still in their infancy. In that revolutionary era, as sacred walls crumbled, almost anything seemed possible—including, for some thinkers, domains beyond the limits of space and time. It is a tale that has grown even more poignant in recent years with the formulation of novel models of unification—beginning with supergravity and superstrings, and leading, most recently, to M-theory and brane worlds. These imagine not just five dimensions, but ten or eleven—extending Kaluza-Klein theory to extraordinary new realms.

Some of the latest approaches offer the tantalizing possibility of physically detecting extra dimensions. Testing Kaluza and Klein's miraculous hypothesis has become an exciting new avenue of experimental physics. Soon we may know if there is more to the world of dimensions than meets the eye.

CHAPTER 1

The Power of Geometry

O WRETCHED race of men, to space confined!
What honour can ye pay to him, whose mind
To that which lies beyond hath penetrated?
The symbols he hath formed shall sound his praise,
And lead him on through unimagined ways
To conquests new, in worlds not yet created . . .

March on, symbolic host! with step sublime,
Up to the flaming bounds of Space and Time!
There pause, until by Dickenson depicted,
In two dimensions, we the form may trace
Of him whose soul, too large for vulgar space,
In n dimensions flourished unrestricted.

—JAMES CLERK MAXWELL to the Committee
of the Cayley Portrait Fund, 1887

Shadow Play

Is the cosmos just a shadow play? Such is its portrayal in the sacred
Indonesian tradition of *Wayang Kulit*. Part religious ritual, part
entertainment, *Wayang Kulit* is a type of puppet theater acted behind
a backlit screen. One of the oldest storytelling traditions in the
world, its nightlong dramas depict the endless struggles of gods and
demons as they set the course of cosmic history.

A typical show begins with the audience seated in front of a
stretched white sheet. An oil lamp bathes the screen in an other-

worldly glow. The Dalang, or puppeteer, takes his place behind the screen and chooses from among two sets of colorful handcrafted leather puppets. One set represents the heroic characters, the other the villains. Behind the Dalang are musicians, whose surreal cadences lend aural texture to the tales. During the course of the performance, the Dalang never addresses the musicians; rather, they shape their sounds around the ever-changing moods of the stories. As they play on, the Dalang conjures up the memories of generations of story-tellers and delivers his one-man epic. From nightfall until the first stretches of the Sun's awakening rays, the consummate puppeteer never takes a break. With his well-practiced repertoire of voices and movements, he evokes the bravery of legendary warriors as they grap-ple with horrific ten-headed monsters, relays the blood feuds of times untold, and sketches the tales of impassioned lovers as they woo and betray each other.

With all eyes gazing intently at the screen, the audience sees only projected images of the backstage drama. Passing through each other, blinking out and then suddenly reappearing, these shadows are able to act in a manner impossible for more solid figures. The specter of a gorgon might easily and instantly devour the projection of a sword-wielding lad, with nary a bulge or burp. Two other creatures might merge their shadows and form a ghastly behemoth. Well aware of the varied laws of the two kingdoms—the colorful one behind the sheet and the murky one on its surface—the Dalang extracts whatever magic he can from the difference.

Strange as it would seem, this exotic fiction could represent the truth—in artifice rather than content. A new movement in physics imagines the universe itself as a shadow theater. The world we see around us, according to this novel vision, is but a mere projection of a more fundamental reality. The true drama takes place beyond the curtain on a higher-dimensional stage. Possessing at least one extra dimension beyond space and and time, this backstage area, called the bulk, can never be seen (with visual means at least) because it admits no light. We can only witness the shadow play on the curtain itself, a three-dimensional volume called the brane, and surmise what lies beyond.

Nevertheless, researchers are trying to test this new model of physics, known as M-theory. Their strategies make use of experiments that rely on gravity rather than light. Like the Dalang and his musicians, gravity is thought to have special access to backstage. According to M-theory, it can penetrate the bulk, and emerge in other places along the brane. If this indeed is the case, then gravity could conceivably jump from one region to another at a rate faster than the speed of light. Physicists are currently using accelerator data and other means in attempts to substantiate such a hypothesis. They are also examining alternative higher-dimensional models of the universe, in various versions of Kaluza-Klein theory.

Plato's Cave

Although M-theory has been around for only about a decade, and more basic Kaluza-Klein theory for less than a century, the notion that the visible world represents mere shadows of the truth is quite ancient. Philosophers have long been intrigued by Plato's idea of "forms." All we see around us, according to the Greek sage, is just an illusion—an incomplete projection of the perfect domain of forms. These forms constitute the ideal versions of everything we know: the perfect Sun, the perfect Moon, the flawless human being, refreshingly pure air and water, and so on.

Plato encapsulated his thoughts on this subject in his famous "Allegory of the Cave." He imagined prisoners constrained to spend their entire lives inside a cave, close to the entrance. Shackles restricted their motions so they could gaze only at a stony wall. From their vantage point, however, they could observe the interplay of shadows from the outside world, cast by a fire blazing outside in the distance. As people carrying all sorts of goods and vessels walked between the fire and the mouth of the cave, the prisoners saw only their silhouettes. Because the captives were unfamiliar with external reality, they presumed that the flat shadows, not the solid bodies, were all that there was.

Viewed in the modern context, Plato's cave allegory seems to imply that our three-dimensional world is but a projection of an even higher-dimensional reality as well. Clearly, though, that wasn't Plato's intention. The ancient Greeks had no known interest in higher dimensions. Plato set his perfect realm in a metaphysical domain, not in a multidimensional extension of our own space. Thus his tale was explicitedly a metaphor for unseen perfection, not for unseen dimensions.

On the contrary, the Greeks saw reason to believe that nature was limited to only three dimensions: length, width, and height. Plato's student Aristotle, born in 384 B.C., emphasized this fact in his work "On the Heavens." Recognizing the natural progression from a line to a plane and then to a solid, Aristotle stressed that nothing of higher dimension lies beyond. He considered the solid to be the most complete type of mathematical object, unable to be augmented or improved. Therefore, no other body could surpass it in number of dimensions. To further bolster his case, he pointed out the Pythagorean idea that three was a special number, because everything has a beginning, a middle, and an end. The Pythagoreans were a learned society in ancient Greece that had a great interest in the power of mathematics and the mystical properties of various numbers. Hundreds of years later, a treatise by Ptolemy entitled *On Dimensionality* amplified Aristotle's thesis. In it Ptolemy showed that one couldn't construct a set of more than three mutually perpendicular lines passing through a single point.

Due perhaps to the hallowed Pythagorean tradition, as further developed by Plato, Aristotle, and others, Greek society maintained a keen fascination with three-dimensional geometry. This extended to its art and architecture, from precisely proportioned sculpture to the grand symmetrical structures of the Parthenon. In mathematics, the Greeks were the first to discover that there are only five regular (equal-sided) three-dimensional polyhedra, known as the Platonic solids. These are the tetrahedron (four-sided pyramid), the cube, the octahedron (eight sides), the dodecahedron (twelve sides), and the icosahedron (twenty sides). This limitation seemed quite mysterious, given that there are an infinite number of regular two-dimensional polygons (triangles, squares, and so forth). Such striking differences between two and three dimensions made the latter seem even holier.

From such rudimentary seeds, Euclid made beautiful structures bloom. He used his postulates to prove virtually all of the basic geometric properties known at the time. Many of those results are familiar to every high school student. For example, if two triangles have exactly the same shape, the sides of one must be proportional to the sides of the other. With an argument based on angles, Euclid also explained why there are only five Platonic solids.

The postulates upon which Euclid based his proofs were so compelling they were considered sacrosanct until the nineteenth century. With the first four, it's clear why. Every set of two points defines a straight line, he noted. A line segment can be extended forever. A circle can be drawn with any given center and radius. All right angles are equal to each other. Who, familiar with simple plane geometry, could argue with these?

Euclid's fifth postulate is markedly more complex than the other four. Consider two straight lines, and a third line crossing them. This creates two intersections. Suppose the angles on one side of both intersections are each less than right angles. Then that is the side where the first two lines eventually meet.

Because of its bearing on the subject of parallel lines, the fifth postulate has come to be known as the parallel postulate. Mathematically it is equivalent to the following alternative statement, called Playfair's axiom: given a line and a single point not on it, there is precisely one line parallel to the first through that point. When expressed in this manner, one can readily see how the parallel postulate serves as a "duplicating machine" for producing parallel lines throughout all of space. If one wants to construct a set of parallel lines, just take one line and choose a point that happens to be somewhere else. The point automatically acts as the basis for a parallel line.

Given the relative elaborateness of the parallel postulate, for generations mathematicians wondered if it could be derived from the other four postulates. In that case it would be a secondary proposition instead of a basic assumption. Euclid himself considered it inferior to his other postulates and, in his proofs, avoided using it as much as he could. Various mathematicians' attempts to dethrone the fifth postulate all failed, however. It wasn't until the early nineteenth century, and the discovery of non-Euclidean geometries, that Gauss, Bolyai, and Lobachevsky demonstrated that the parallel postulate was wholly independent of the others, and could in fact be replaced with other assumptions.

Until then, Euclid's *Elements* reigned supreme in the field of geometry. It is a record thus far unsurpassed by any other scientific work, and a tribute to the magnificence of Greek thinking on the subject.

Leonardo's Perspective

When Rome conquered Greece, it acquired a cargo of natural and philosophical knowledge, which it unpacked and wore with great enthusiasm. Though the Romans had many erudite thinkers, much of their scholarship came secondhand. They had little interest in developing their own theories. Still the older ideas were none the worse for wear, and helped them construct magnificent temples, statuary, and other public works, with designs derived from Greek mathematical principles.

The rise of Christianity and the fall of Rome led to a radical change of attitude in Europe toward science and culture. The extravagance of Greco-Roman art and architecture became replaced by austerity and uniformity. Thoughts turned to preparations for the world to come, rather than ways to understand the world that is.

Throughout the Middle Ages, a period dating roughly from the fifth until the fourteenth centuries, an emphasis on unadorned design resulted in a two-dimensional approach to painting. Portraits from that era appear flat and unrealistic, like paper dolls. The notion of depth was almost forgotten, as painters reproduced staid likenesses of Jesus, Mary, the Apostles, and other New Testament figures.

Then in the Renaissance era, the sleeping giant of creative art arose from its slumber. Rubbing its eyes, it gazed at the world anew. It began to scrutinize the precise details of the way nature appears, capturing those impressions in increasingly realistic depictions.

One of the harbingers of the new movement was the early fourteenth-century Florentine artist Giotto di Bondone. When Giotto painted scenes, he imagined them from the point of view of someone standing a certain distance away. Then he sketched the images with those lines of sight in mind. The result was sharply different from the flat paintings of his predecessors, far more vivid and true to form.

With his discovery of perspective, Giotto brought the third dimension back into art. Onlookers stood entranced when looking at Giotto's paintings, like children watching television for the first time. They marveled at his ability to make them feel as if they were actually at the scenes he rendered. Soon other artists began to imitate his style, hoping to create some of their own striking images.

To improve upon the illusion of three-dimensionality, artists began to study the long-ignored field of Euclidean geometry. They also began to realize that the proper placement of light and shadows would enhance the realism of their portrayals. Thus, the best artists also became naturalists and mathematicians, calculating the best color and placement for every aspect of their works.

Perhaps the quintessential Renaissance artist was Leonardo da Vinci. Leonardo, who worked in the late fifteenth and early sixteenth centuries, was determined to render his portraits as lifelike as possible. To turn his canvas into a mirror of nature, he studied mathematics, mechanics, optics, anatomy, and other scientific subjects, exploring them in groundbreaking ways. His notebooks contain some of the most detailed studies of human and animal forms ever rendered, indicating the precise exertions of various muscles in a variety of movements. These sketches helped him create realistic portrayals of his subjects that almost seem to be gazing, or even smiling (in his best-known masterpiece), back at the beholder.

Leonardo was very interested in the subtle dance of light and shadow that the Sun's rays produce on a subject. He noticed that lighter and darker areas can be used to convey a sense of either proximity or remoteness. By mixing his colors with appropriately sunny or dusky shades, he found he could enhance the three-dimensionality of his works.

Corresponding to the development of depth in painting came a revived interest in sculpture. Even more so than its Greek and Roman antecedents, Renaissance sculpture captured the flesh and blood humanity of its models. Michelangelo's towering *David,* with its youthful strength and resolute expression, is perhaps the finest example of this movement.

A Star to Steer By

We now know the universe to be immeasurably large, perhaps even infinite in its extent. Until modern times, however, humankind was unaware of this vastness of scope. The medievals, for instance, envisioned the world as a flat plate, surrounded by a relatively nearby dome of celestial bodies. In their perspective, earthly existence was literally confined to a plane, with all above considered spiritual, and all below, demonic.

The expansion of artistic space in Renaissance Europe, with the introduction of perspective and the creative use of light, was accompanied by a rethinking of the magnitude of physical space as well. During the Age of Exploration, which spanned the late fifteenth and the sixteenth centuries, intrepid sailors such as Columbus and Magellan established the roundness of our planet. Then in the early seventeenth century, pioneering astronomer Galileo Galilei surveyed Venus, Jupiter, the Moon, and other objects through his telescope. He established that many celestial bodies have shared features. Jupiter has moons, he showed, just like Earth. Venus has phases, he demonstrated, just like the Moon. And the Moon has mountains almost terrestrial in appearance. From such evidence, Galileo concluded that the Earth and the other planets have equal standing in the solar system, each revolving around the Sun. This supported the Copernican view, advanced in 1543, of a Sun-centered solar system.

Galileo also aimed his telescope at the stars. Even magnified, he noticed that they still remained points rather than disks. For this reason, he surmised that they are far more remote than the planets. Turning his instrument to the haze of the Milky Way, he showed that it harbors vast quantities of stars. Innumerable bodies of light speckle the farthest reaches of the heavens, he inferred.

Galileo's discoveries led to increased speculation that the universe was unlimited in size. In 1686, French scientist Bernard Fontenelle in his treatise *Conversations on the Plurality on Worlds* contemplated the possibility that the distant stars have orbiting planets of their own. Could space keep on going forever, full of endless possibilities? Such thoughts echoed the writings of Neapolitan monk Giordano Bruno, who, even before Galileo's findings, had argued on philosophical grounds that the universe contains an infinite variety of worlds.

Around the same time Fontenelle's book appeared, English physicist Isaac Newton published the *Principia*, an enormously influential description of the laws of motion and gravity. Newton's cosmic epic is set on a vast, three-dimensional stage in which celestial *dramatis personae* such as stars, planets, moons, and comets interact with one another by means of gravitational forces. Motions take place along three perpendicular axes—x, y, and z—imagined to extend throughout all of space in infinite straight lines. Newtonian physics also includes a fourth axis, called *t,* that represents the time particular occurrences take place. Thus, as Einstein and Infeld pointed out in their book *The Evolution of Physics,* the *Principia* contains the seeds of the notion that the universe is four-dimensional, with the fourth dimension representing duration.

It's about Time

It is a commonplace belief that Einsteinian relativity, developed in the early twentieth century, ushered in the concept of a four-dimensional cosmos. In truth, the idea of time as the fourth dimension dates back much further than that. The fact that H. G. Wells in his 1895 novella *The Time Machine* wrote that "there are really four dimensions, three which we call the three planes of Space, and a fourth, Time,"[1] suggests earlier interest in the subject.

The first documented reference to the fourth dimension is in an article, "Dimension," written in 1754 by the French mathematician Jean d'Alembert. D'Alembert was an influential interpreter of Newtonian physics who had written considerably on the dynamics of objects. The piece appeared in the well-respected *Encyclopédie,* an alphabetical compendium of the knowledge of the times, edited by Diderot and d'Alembert. In his article, d'Alembert discussed the idea of solids having length, width, and height. He then imagined combining the three spatial dimensions with the direction of time to form a four-dimensional whole.

We cannot give d'Alembert full credit for the discovery of the fourth dimension, however. Mysteriously, he ceded the credit to another scholar, one whom history has yet to reveal. He referred to the unknown author as an "enlightened man he knows," who stated that it is "possible to conceive of more than three dimensions . . . by regarding duration as a fourth dimension."[2]

Who was this mystery man? Some scholars think it was the brilliant French-Italian mathematician Joseph-Louis Lagrange, who later kept up a steady personal correspondence with d'Alembert.[3] Indeed, in 1797, Lagrange wrote the second known reference to the fourth dimension in his text *The Theory of Analytical Functions.*

In it, Lagrange reports that one can consider mechanics as a four-dimensional geometry, with $x, y, z,$ and t as the four coordinates.

The main problem with the hypothesis of Lagrange as the originator is that he was only eighteen at the time of the publication of the *Encyclopédie*. Although he was precocious, it would have been highly unusual for such an extremely young thinker's ideas to be cited with such authority. On the other hand, Lagrange did become a professor the following year, making some important contributions to calculus and establishing his credentials as an innovator in his field. Perhaps we will never know whether Lagrange, d'Alembert, or another insightful mathematician was the first to identify the remarkable juxtaposition of space and time that has proven essential to modern science.

And God Created Space

Time is one way of envisioning the fourth dimension. Another way to conceive of it is to ponder more than three spatial axes. Could there indeed be a fourth *spatial* dimension? At first glance, Ptolemy's argument would seem to preclude such a possibility. One cannot draw a line at right angles to a set of three perpendicular axes. No matter

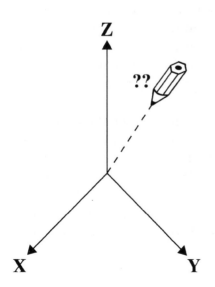

No matter how hard one tries, one cannot draw a line perpendicular to all three spatial axes.

how hard one tries, the fourth axis would end up in a nonperpendicular direction relative to at least one of the others.

Yet could such an inability to construct a fourth spatial dimension be a limitation of our own perception rather than a restriction on space itself? Could bizarre beings in other realms have the ability to perceive four, five, or even higher numbers of dimensions? Other animals have heightened senses we don't have, such as dogs' ability to hear high-pitched whistles. Why couldn't the same be true, somewhere in the universe of possibilities, for the sense of dimension?

The German philosopher Immanuel Kant was the first known thinker to grapple with such possibilities. In his first published treatise, *Thoughts on the True Estimation of Living Forces,* which appeared in 1747, he considered what it is about the world that makes us observe it as three-dimensional. Drawing on Newtonian theory, he hypothesized that the answer has to do with the nature of forces. Objects interact with each other by means of gravity, he noted, which varies inversely with the squares of the their mutual distances. The mathematical form of this law mandates that space has three dimensions.

How could the law of gravity tell us anything about the dimensionality of space? Consider the relationship between the gravitational pulls of two imaginary planets, both with the same mass, but with the first having twice the radius of the second. Then, by Newton's inverse-square law of gravity, a boulder resting on the first planet should be only one-fourth as heavy as that of an identical boulder lying on the second. This result depends, however, on the planets being three-dimensional. If they were two-, four-, or six-dimensional, the boulders' weights would have altogether different relationships. Hence the precise way gravity drops off with distance relates to the number of spatial dimensions.

Kant then contemplated the likelihood of other realms, with different laws and varied numbers of dimensions. Because God can create anything, he surmised, we must allow for such a possibility. If these other realms exist, though, we have yet to see them. Therefore, they must be totally detached from our own world, beyond all prospects of contact. Like Australia and New Guinea before Europeans arrived, they would be isolated islands with vastly different conditions. As Kant described:

If it is possible that there are extensions with other dimensions, it is also possible that God has somewhere brought them into being; for His works have all the magnitude and manifoldness of which they are capable. Spaces of this kind, however, cannot stand in connection with those of a quite different constitution. Accordingly such spaces would not belong to our world, but must form separate worlds.[4]

After considering such strange, disconnected domains, Kant used theological arguments to demonstrate their unlikelihood. God prefers harmony, he argued, not discord. Therefore, he concluded, although it is wise to consider all possibilities, God probably created the universe as a unified three-dimensional entity, rather than as a jumble of various dimensional worlds.

Through the Looking Glass

Later in his life, Kant became intrigued by the notion of chirality, or handedness. Why are some things right-handed and others left-handed, and how can one unambiguously describe the difference? For example, there are doors that open on the right, and doors that open on the left. There are screws that turn clockwise, and ones that turn counterclockwise. Is there some unequivocal way of telling them apart, without recourse to directional terms such as "right," "left," "clockwise," and "counterclockwise?"

The question of identifying handedness is deeper than it would seem at first. Suppose you are in radio contact with an alien, an amorphous creature named Xyl who is unfamiliar with the human form. You wish to convey to Xyl the fact that you write with your left hand. You try to describe the difference between left and right hands, but Xyl is confused. You can't say "when I hold a pen the thumb is on the right" because Xyl doesn't understand what "right" is. You can't even resort to directions, such as "when you face due north, your left side points west," because Xyl doesn't know terms such as "north" and "west" either. Perhaps if you are clever you can make your point by referring to some scientific phenomena Xyl is familiar with, such as the relative location of particularly bright stars in a distant galaxy. However, conveying this information is much more cumbersome, than, say, explaining that you have five fingers on each hand.

It is a curious feature of nature that it possesses both right- and left-handed things. Scientists used to believe that on a fundamental level there were equal numbers of each, until they discovered that certain particle interactions tend to favor one over the other. Now we know that nature, though offering examples of both types of chirality, does indeed have its preferences.

Could a right-handed thing ever be turned into its left-handed equivalent? Clearly yes in the case of flat objects. If you flip over the

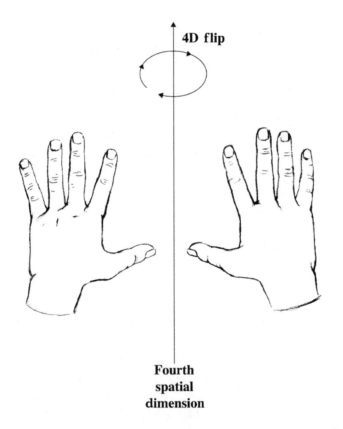

4D flip

**Fourth
spatial
dimension**

*A left hand can be transformed into a right hand only by flipping
it through a fourth spatial dimension.*

Roman letter *R*, you easily produce something that resembles the Cyrillic letter Я. Note, though, that the flipping over must take place outside the plane in which the *R* is located. Therefore, although the *R* is a two-dimensional object, it must be lifted up and turned over in the third dimension.

Similarly, as the German mathematician August Möbius first pointed out in 1827, three-dimensional bodies could be converted from right- to left-handed only by flipping them in a hypothetical fourth spatial dimension. No rotation confined to ordinary space could enable them to change their chirality.

For example, suppose you own a shoe manufacturing plant, and are aiming to cut costs. You wish to find a means of converting right shoes into left shoes, so you only have to produce one type. Try as you might to rearrange the shoes without ruining them, barring access to a fourth spatial dimension, you'd find your task impossible.

Interestingly, Möbius, much later in his life, fashioned a simple way of flipping two-dimensional objects. Called the Möbius strip, it consists of a thin band of paper, twisted half a turn before the two ends are glued together. (It was independently discovered by Johann Listing, a few months before Möbius, but somehow Möbius gets the credit.) Because of the half-twist, the strip has only one side—unlike, for instance, a buckled belt. This can be verified by taking a pencil and drawing a continuous line that visits both the "inside" and the "outside" of the strip. Both the inside and the outside turn out to be the same.

Now cut out a piece of paper to look like an *R* and run it along the outside of the strip. As the outside turns into the inside, the *R* flips over and becomes a Cyrillic Я. Hence the Möbius strip acts as a chirality-changing machine!

The equivalent of using a Möbius strip to flip two-dimensional objects is to use something twisted in the fourth spatial dimension to flip three-dimensional objects. One version of such a monstrosity is called the Klein bottle. It looks like a vase in which the neck is elongated, twisted inside out and reconnected with one of the sides. In this manner, the inside of the vase corresponds to the outside. Does such a construct hold water? Only if a four-dimensional twist could somehow be achieved would topologists ever know.

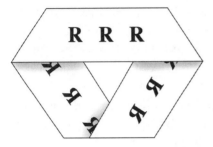

The Möbius strip, a band given a twist and attached to itself in a loop, can invert two-dimensional objects. Here it reverses the eighteenth letter of the Latin alphabet to become the final letter of the Cyrillic alphabet.

Challenging Euclid

By considering the mere possibility of higher spatial dimensions, Kant and Möbius each raised interesting philosophical questions. Does the diversity of the cosmos allow for worlds of varied dimensionality in regions separate from ours? Does the presence of left- and right-handed objects imply a fourth-dimensional transformation that converts one to the other?

Despite their curiosity, neither Kant nor Möbius attempted to explore the properties of the extra dimensions they hypothesized. In Kant's case such disinterest derived from his belief that such domains, if they existed, would remain forever disconnected from our own world, and hence beyond our understanding. Our own world, he later emphasized in his famous "A Critique of Pure Reason," is a three-dimensional realm, fully described by Euclid's laws. Kant believed that Euclidean geometry derived from "pure intuitive knowledge" and was therefore an absolute description of the way things are. Therefore, in Kant's view, trying to disprove Euclid would be tantamount to saying that red is green or that one plus one equals three. Kant's adamant perspective, influential even beyond his lifetime (he died in 1804), stood in marked contrast to the more skeptical outlook of his contemporary d'Alembert. D'Alembert called the problems with the parallel postulate "the scandal of elementary geometry."

By the time of Möbius, who began his studies shortly after the era of Kant, mathematicians were starting to chip away at the Euclidean monolith. One of Möbius's professors, Carl Friedrich Gauss, was among the first to envision non-Euclidean geometries. Intimidated, though, by the residual hold of Kantian thought, Gauss was quite circumspect in his beliefs. To avoid controversy, Gauss never published his ideas on the subject. Möbius, whose thesis was in astronomy and who went on to be a pioneer in the field of topology, was therefore very likely unfamiliar with Gauss's non-Euclidean ventures.

Despite his cautious attitude, Gauss was pleased when a much later student of his, Georg Friedrich Bernhard Riemann, developed a brilliant new non-Euclidean approach. Not only did Riemann's novel perspective allow for alternatives to Euclid's postulates, it also addressed the structure of higher-dimensional spaces, called hyperspace. Although Riemann's inaugural lecture on this topic took place in 1854, it wasn't published until 1868, the year Möbius died. Only then could the mathematical exploration of higher-dimensional realms begin in earnest.

Not only did Riemann's stunning breakthrough radically change perceptions of higher spatial dimensions, it also paved the way for Einstein's general theory of relativity. Moreover, it cleared vast new territories upon which a host of unified field theories—from Kaluza's onward—would eventually settle. Today's theoretical physicists are well aware of the debt they owe to Gauss's prize student.

Can an entire town take credit for a discovery? Sometimes the place where a researcher works is just incidental. However, in the case of Göttingen, the cozy German town where Gauss reigned, Riemann triumphed, and numerous other mathematicians later made pivotal contributions, it's clear that a certain bewitching spirit in the air must have played at least some part.

CHAPTER 2

Visions of Hyperspace

> *To the question, where and what is the fourth dimension, the answer must be, it is here—in us, and all about us—in a direction toward which we can never point because at right angles to all the directions we know. Our space cannot contain it, because it contains our space. No walls separate us from this demesne, not even the walls of our fleshy prison; yet we may not enter, even though we are already "there." It is the place of dreams, of living dead men. It is At the Back of the North Wind and Behind the Looking Glass.*
>
> —CLAUDE BRAGDON, *Four-Dimensional Vistas,* 1916

Season of the Witch

The snow lingers late in the Harz Mountains in central Germany. It clings to her crags and peaks like a long white cloak concealing a secret beauty. Then in springtime, the mystery is revealed. Beneath the warm sun, magnificent forests burst forth in greenery, inviting villagers to wander through them along serpentine paths. Half-timbered houses bask in welcome glows, as flowers poke their heads from long wooden pots.

Just around the time that spring breaks through, there is an ancient Harz tradition that marks the border between darkness and light. It is Walpurgis Night, the festival of demons, witchcraft, and fertility. In each little village, shops display handcrafted witches, supposed to bring good luck. Legend has it that on April 30 each year, a band of particularly powerful lady conjurers gather on the Brocken, the highest peak in the Harz, to dance away the snow and bring on new life. They serve as assistants to Walpurga, goddess of cultivation. A horrific scene in Goethe's *Faust*, Mussorgsky's eerie composition "Night on Bald Mountain," and the spine-tingling finale of Disney's *Fantasia* each capture this hallowed tradition.

In the foothills of the mountains lies a center of mystery and logic, of facts and folklore, that is renowned for its scholarship. Legendary Göttingen is one of the oldest and most prestigious university towns in Europe, and has been home over the years to more than forty Nobel laureates. Surrounded by a devilishly jagged landscape, it has always been isolated from the major centers for industry and commerce. Yet its product, education, has proven priceless for the history of science.

The most famous scholars of Göttingen, whose faces used to appear on German banknotes, reflect its dual character. Jakob and Wilhelm, the Brothers Grimm, captured the mythos of the mountains and forests in their riveting tales. They set down centuries of

A street in mathematical Göttingen, one of Germany's oldest and most respected university towns.

folklore in some of the most widely read collections ever written. Then there was Gauss, the legendary mathematical genius. Gauss and his successors transformed the mission of Göttingen from primarily a German-focused enclave for the study of religion, literature, and culture to an internationally recognized leader in mathematics and science. That reputation continued for more than a century.

Göttingen has not always been kind to its heroes. Its cobbled streets and half-timbered houses will always bear the mark of two dark days in its history. Through those streets and past those houses, two witch hunts separated by a hundred years in time forced men of greatness to flee in terror. In 1837, the Brothers Grimm and five others, known as the Göttingen Seven, were dismissed from the university because of their opposition to a reactionary constitution established by the Kingdom of Hannover. At that time Göttingen was part of that kingdom, before Hannover joined a unified Germany. Then, in the mid-1930s, the mathematics department was decimated when Nazi racial laws forced most of the professors out, including the director, Richard Courant. Göttingen mathematics would never fully recover. In terms of Göttingen's prestige, the great age of Gauss would finally be over.

The Humble Genius

Carl Friedrich Gauss, an extremely modest man, would have been surprised that his influence lasted so long. He was born in 1777 to a poor family in Brunswick (Braunschweig), Germany. His father, Gerhard, was a gardener and bricklayer; his paternal grandfather, a peasant. His mother's side of the family were stonecutters and weavers. If it weren't for a series of fortuitous circumstances, he almost certainly would have ended up in one of the family trades.

Even before the age of three, it was clear that Gauss had miraculous powers of computation. One day, Gauss's father was completing the payroll for his employees. Gauss, just a toddler, watched his dad attentively. Suddenly he called out, "Father, your calculations are wrong. It should be . . ."[1] The boy's figure was absolutely correct. Somehow, at an age in which he was barely able to talk, he had taught himself arithmetic.

Carl Friedrich Gauss, the foremost early nineteenth-century German mathematician, was one of the founders of non-Euclidean geometry.

As impressed as Gerhard was, he still wanted his son to be a tradesman. Such talents would best be used for practical purposes, he felt. He never understood why anyone would want to pursue mathematics for its own sake. Every chance he could, he would lay obstacles in his son's path toward an education.

Fortunately, Gauss could count on the strong support of his mother, Dorothea. She defended her son's interests whenever humanly possible, protecting him from the shortsightedness of her husband. Gauss was also lucky to have a perceptive uncle who nourished his talents. Uncle Friedrich saw in Gauss's bright young eyes the gift of genius. He fed the boy a steady diet of commentary on various intellectual subjects, which the precocious youth ravenously consumed.

There is no shortage of stories about the young thinker's achievements. When Gauss went to school, he impressed his first teacher so much with his mathematical prowess (he instantly added a series of one hundred numbers in his head) that the teacher bought him the best arithmetic textbook around. Much to the teacher's surprise, even that book was too easy for him. By the age of twelve, Gauss had already started thinking about alternatives to Euclidean geometry. Influenced, no doubt, by his uncle's inquisitive spirit, he took absolutely nothing for granted.

When Gauss turned fourteen, another of his teachers, the mathematician Johann Bartels, introduced him to a powerful figure, Carl

Wilhelm Ferdinand, Duke of Brunswick. Impressed with his modesty and stunned by his talent, Ferdinand offered to sponsor Gauss's education. He paid for Gauss to attend the University of Göttingen, which the young genius began at age eighteen.

While a student at Göttingen, Gauss found a lifelong friend in Wolfgang Bolyai, who was also studying math. One day, he brought Bolyai home to meet his mother. Dorothea Gauss took Bolyai aside and asked him if he thought her son would ever amount to anything. Bolyai confidently replied, "Only the greatest mathematician in Europe,"[2] bringing Dorothea to tears.

Bolyai comforted Gauss in trying times. In 1806, after Gauss left Göttingen, received his doctorate from Helmstedt, and returned to Brunswick, France launched an invasion of German lands. A valiant soldier as well as a statesman, Ferdinand unsuccessfully took up arms to defend his native soil against Napoleon. After the duke was mortally wounded, Gauss watched in horror as his dying patron's carriage passed his house. Gauss knew that from that point on, he was master of his own fate.

Though Napoleon robbed Gauss of his mentor, the French general's policies did not hinder his career. Thanks to the influence of the naturalist Alexander von Humboldt, a German then living in Paris, Gauss was appointed professor at Göttingen and director of its observatory. A few years later, the French were driven out and the Kingdom of Hannover restored. By then, Gauss was well settled and immersed in his exceptionally productive research endeavors, including his work in non-Euclidean geometry.

Strange Parallels

It is a quirk of scientific history that often two or three researchers independently and simultaneously arrive at the same results. In many cases, the times are just right for a solution to be found. Therefore, perhaps it is not surprising that non-Euclidean geometry had three authors, each working separately.

One of them was Gauss, who discreetly examined the problem without ever publishing his solution. Another was Wolfgang Bolyai's son, János. While teaching him mathematics, Wolfgang warned János not to waste his time investigating the issue of Euclid's parallel postulate. Like most curious sons would in those circumstances, János did exactly the opposite. He made a full study of the question, coming to the conclusion that the parallel postulate was independent of the other four. He found that he could replace it with an alternative supposition. In 1823, he excitedly reported to his father that he had discovered a strange new world. In this topsy-turvy kingdom, instead of just one, there can be an infinite number of lines, each passing through the same point and parallel to another line. In other words, a line and a point can define an unlimited set of parallels. Moreover, instead of triangles having exactly 180 degrees, in this world they always have less. Several years later, János's father helped him publish the extraordinary results, sending an advanced copy to Gauss.

When Gauss read over the report, he delivered both a boon and a blow to young János's ego. Gauss told him he was a genius, while at the same time informing him that he wasn't the first to develop a non-Euclidean geometry. For the first time, Gauss revealed the full extent of his own studies on the subject. János took the bad news hard. He never published again in the field of mathematics. Instead, he spent the rest of his career in the Austrian army, where he was known as an expert dancer and swordsman.

The third Amundsen of the non-Euclidean terrain was the Russian mathematician Nikolai Lobachevsky. Like Gauss, Lobachevsky had been a student of Bartels, who introduced him to the parallel postulate dilemma. Energized by these discussions, Lobachevsky set aside the postulate, like an Arctic explorer leaving behind excess baggage. He then constructed a geometry similar to János Bolyai's. Neither Bolyai nor Lobachevsky were familiar at first with each other's work. This is not surprising considering that Lobachevsky published in an obscure Russian journal called the *Kazan Messenger*. Eventually, Gauss found out about Lobachevsky's paper, praised it, and recommended him as a corresponding member of the Royal Society of Sciences in Göttingen. Gauss's notes on his own non-Euclidean research, along with his personal copies of the works of Bolyai and Lobachevsky, were discovered only posthumously in the mid-1850s, and made widely available only in the late 1860s.[3]

Compared to the ancient Greek vision, non-Euclidean geometry is as twisty as the roads leading into Göttingen. If the Brothers

Grimm were asked to describe their colleague's work, perhaps they
would set it as a legend in the most disorienting forest imaginable:

> On an unusually chilly evening in late April, a woodcutter and his
> son were walking home through the thick woods. Each was looking
> for stray timber they had chopped that morning, so they decided to
> take separate paths. Because it was so easy to get lost in the forest,
> the woodcutter asked his son to follow a trail parallel to his.
>
> While the son was adjusting his lantern, he accidentally
> stepped on a flower that was planted by Walpurga herself. Enraged
> by the loss, Walpurga consulted her witches, who set a curse upon
> the land. From where the boy stood, instead of one path parallel to
> his father's, they created an infinite number of parallel trails. Just
> then, the woodcutter called out, "Where are you son? Are you still
> journeying in the same direction as me?" But the boy, transfixed by
> possibilities as mesmerizing and multifarious as the myriad stars
> above him, could go no farther. And it is whispered in the hills that
> his father never found him.

The Trials of a Preacher's Son

Non-Euclidean spaces can divide, but they can also unite. It was late
in life that Gauss met his true disciple, a shy young student who was
nevertheless courageous enough to complete the old professor's
dream. Through the work of his student, Gauss would live to see a
secret goal fulfilled: the rebuilding of geometry on a sturdier foundation.

Gauss first encountered Bernhard Riemann, the son of a
Lutheran minister, during a class he taught in basic mathematics.
The nineteen-year-old Riemann had arrived in Göttingen in 1846,
fully intending to study theology. He hoped to follow in the footsteps
of his father. Once he began university, however, Riemann realized
that mathematics was his true calling. After begging his father for
permission, he switched his field of study and started attending lectures by Gauss and others.

Riemann was nervous, no doubt, as he took his place in the lecture hall. Before entering, he had to show proof that he had paid for
the course. In those days, each professor charged an individual fee
for the privilege of a seat. Gauss set his admission price particularly

high because he greatly preferred research over teaching and liked to keep his classes small.

Like Gauss, Riemann came from a poor family. It was a blessing for him that he could even pay for his classes. Malnourished as a child, he was plagued by ill health for much of his life. Moreover, he suffered painful anxiety in social situations, feeling truly comfortable only with his closest friends and relatives. Therefore, although Riemann greatly admired Gauss, he was intimidated by his presence.

Riemann was not fearful, however, of the material that Gauss presented. He was as confident with equations as he was awkward with people. After mastering his courses with ease, he left Göttingen in 1847 for the University of Berlin, where he studied with several prominent professors. After two years, he returned to Göttingen and began to broaden his perspective with several courses in philosophy and physics.

In 1850, Riemann became fascinated by the work of the experimental physicist Wilhelm Weber. Weber, one of the Göttingen Seven who had recently been allowed to return, was an expert in the behavior of electricity and magnetism. He was a friend of Gauss and had consulted with him about these subjects.

Through familiarity with Weber's studies, Riemann became struck by the idea of devising a single mathematical theory for all the laws of nature. He wondered if there might be a common way of explaining gravitation, electricity, magnetism, thermostatics, and other physical phenomena by use of a solitary principle. Around the same time, he attended lectures in the field of topology by Johann Listing, the true inventor of the Möbius strip. Enriched by knowledge of both abstract spatial relationships and tangible physical properties, Riemann began to envision ways he could use the distortion of geometry to describe all of physics. In this manner, he anticipated the interests of Einstein, Kaluza, and others by more than half a century.

By that time, Riemann was well on track to completing his doctorate and going on to be a professor in pure mathematics. For a student at that stage, one of the most stressful things to happen is becoming engrossed in a completely independent research project. Yet that is exactly what became of Riemann. As more and more of his time became spent in mathematical physics in addition to his normal thesis work, he must have felt like a skier heading toward a tree with feet on either side.

In Germany, the process of obtaining one's degree and becoming a lecturer takes multiple steps. First, one needs to complete a dissertation. Riemann did this in 1851, submitting a masterful treatise on complex variables that thoroughly impressed Gauss. The next stage is a public defense of the thesis, which went well for Riemann, followed by another paper called the *Habilitationschrift,* or probationary essay. Because of his dual interests in pure mathematics and mathematical physics, this took him much longer than usual. After finally submitting his paper, Riemann proceeded to the ultimate step, the trial lecture.

Some students are a trifle apprehensive at that stage. For Riemann it was like plunging into a dark abyss. A number of factors contributed to a temporary nervous breakdown. First, his heart was in physics, not in potential lecture topics. Most of his time was consumed by researching that subject as well as by working as Weber's teaching assistant. Second, at that point Gauss was seriously ill. Riemann feared that he would die before granting him the critical permission to teach. Finally, Riemann himself was not in the best of health, plagued by weak stamina. As he described that dark period:

> I became so absorbed in my investigation of the unity of all physical laws that when the subject of the trial lecture was given me, I could not tear away from my research. Then, partly as a result of brooding on it, partly from staying indoors too much in this vile weather, I fell ill; my old troubles recurred with great pertinacity and I could not get on with my work.[4]

It was June 10, 1854: the day hyperspace would be born. Gauss was well enough that day to administer the trial lecture, so it was to proceed. Riemann mustered up what strength he could, somehow pulled himself together, and reported to the lecture room. Following procedure, he had prepared three possible topics, which he had submitted to Gauss ranked by how familiar he was with each. To Riemann's amazement, Gauss broke with tradition and slyly selected his third topic: "On the hypotheses which lie at the foundations of geometry."

When Riemann started speaking, he had the suspicious feeling that Gauss knew an awful lot about this subject. It was like giving a lecture on cultivating flowers to someone with a lifelong secret garden. With each word, Gauss smiled and nodded in agreement. With bubbling enthusiasm he applauded Riemann's talk and granted him permission to begin his lectureship.

Riemann served for many years as a privatdozent in Göttingen, and then as a professor. All the while he tried in vain to describe all of physics by means of geometry. Unfortunately, because his medium was space rather than space-time, his project was doomed to failure. Nevertheless, his work would later serve as the structural framework with which Einstein would construct his marvelous edifice of general relativity.

At the age of thirty-nine, Riemann developed tuberculosis. To attempt a cure, he traveled to the Italian village of Selasca, on the shores of Lake Maggiore. He died in the summer of 1866, eleven years after the death of his mentor.

Blueprint for Hyperspace

Riemann's inaugural talk was published in 1868, around the time that the non-Euclidean writings of Bolyai and Lobachevsky first became widely available. This triple punch had a lethal impact upon the two-thousand-year-old dotard of Euclidean mathematics. The old outlook was dead; long live Riemannian geometry.

The Riemannian perspective replaced the rigid idea of compass-drawn lines, planes, and shapes with the more flexible concept of a *manifold*. Briefly put, a manifold is a collection of points each characterized by a set of numbers, known as its coordinates. If the manifold is two- or three-dimensional, then each point has two or three coordinates, respectively. However, because the possible number of coordinates is unlimited, one can construct a manifold with any number of dimensions. This means that one can describe a hyperspace of four dimensions as easily as an ordinary space of three dimensions.

A second novel concept in Riemannian geometry is the notion of a *metric*. A metric generalizes the Pythagorean theorem, the mantra of high school geometry teachers, by providing a more malleable way of defining distances between pairs of points. In conventional Euclidean space, the Pythagorean theorem specifies the square of the distance to be the sum of the squares of the x coordinate difference, the y coordinate difference, and so forth. The famous "three-four-five" right triangle is a prime example of this. In Riemannian space, on the other hand, one can alter the distance formula by designing different metrics. One could, for instance, define the square of the distance to be twice the square of the x coordinate difference plus seven times the square of the y coordinate difference, and so on. All one would have to do is construct a new metric to that effect. Thus yardsticks can be stretched or compressed in as limber a fashion as in a Dali painting. This offers the possibility of an unlimited array of structures, most of which would have been unrecognizable to Euclid and his followers.

Two other notions, called *curvature* and *embedding*, further illustrate the differences between Riemann's construct and its antecedents. Euclid's world contains curved lines, such as an arc of a circle, and curved surfaces, such as a section of a sphere, but no such thing as a curved space. For a Riemannian geometry, in contrast, any region of any dimensional manifold can be curved. Thus a three-dimensional volume can just as well be warped as an old vinyl record. A natural question in the case of three- or higher-dimensional spaces is, warped into what? One way to address such an issue is to embed (implant) the curved manifold in a space of even greater dimensions. Thus one can well understand the curvature of a two-dimensional spherical surface by considering it embedded in a three-dimensional space. Similarly the curvature of a hypersphere (the three-dimensional equivalent of a spherical surface) expresses itself nicely when it is housed in a four-dimensional space. The curvature and metric are mathematically related concepts; each can be derived from the other. That is, by defining the distances between all the points in a manifold, one can determine the structure of its curvature, and vice versa.

Riemann's recasting of geometry permits non-Euclidean possibilities even stranger than those of Bolyai and Lobachevsky. For example, in some situations, lines are finite rather than infinite, and possess no parallels to themselves. This violates fully two out of five of Euclid's postulates, sufficient for him to double-flip in his grave!

(The judges on Olympus reportedly awarded him a medal for post-humous reactive gymnastics.)

One can see this violation in the case of spherical surfaces. In Riemannian parlance they are said to have "positive curvature," meaning they are curved around a central point. Take, for example, an orange about to be sliced. Consider the paths a knife would take, from the orange's "north pole" to its "south pole," to be its lines. Beginning and ending at points, clearly these lines cannot be extended indefinitely. Moreover, because all lines eventually meet, none can be said to be parallel.

Interestingly, if one takes an orange slice, then cuts it along its "equator" one can make a triangle. Adding up the angles of the triangle, one finds the sum to be greater than 180 degrees. Contrast this with the Bolyai/Lobachevsky geometry, called "negative" curvature, where each line and external point define an infinite array of parallels, and the sum of the angles of a triangle always yields less than 180 degrees. Strange new worlds indeed.

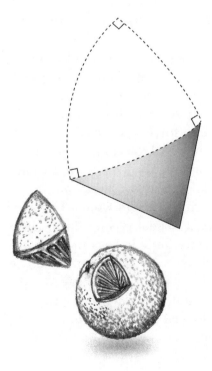

In a positively curved form of non-Euclidean geometry, the angles of a triangular orange slice add up to more than 180 degrees.

Truly, What's the Matter?

Unlike pure, abstract spaces, the physical universe contains matter. It harbors myriad objects, interacting with one another in a complex web of relationships. Massive bodies—from rings spinning around Saturn to apples falling toward Earth—tug on one another by means of gravity. Electrical charges embrace or reject their kin, based on their respective signs (plus or minus, that is, not astrological).

Newton characterized such pushes and pulls as actions at a distance, like invisible chains connecting the mechanisms of a universal machine. By the time of Gauss and Riemann, however, physicists began to envision quite a different explanation, one in which special media, called fields, produce the effects of forces by reacting to the effects of one object and conveying this disturbance to another. Gauss himself developed a mathematical way of discerning the effects of electrical charges on field lines known, appropriately enough, as Gauss's law. If the action at a distance view can be described as one tugboat pulling another along with a rope, the field perspective imagines the first disturbing the water and causing the other to rock in its wake.

Once Riemann mapped out the structure of non-Euclidean geometry, this structure provided a natural way of envisioning such ripples. Instead of thinking of fields as independent entities within space, could they be part of the fabric of space itself? Then could spatial geometry serve as the conduit for force? Riemann's obsession with such a possibility—to fulfill his goal of uniting physics—wracked his nerves and ended in failure. Yet it inspired another mathematician to take even bolder steps in such a direction.

William Kingdon Clifford was born in 1845 in Exeter, England. He had an uncanny gift for visualizing spatial relationships. Once when he was a boy, a family acquaintance brought back from India a challenging three-dimensional puzzle, consisting of a sphere made of intricate interlocking pieces. The goal was to find a way to separate all the pieces, one by one. During dinnertime, young Clifford was shown the puzzle. He diligently looked it over without touching it, then thought over the situation for a few minutes. He picked up the puzzle and instantly solved it.[5]

A prize-winning student, Clifford excelled at both King's College, London, and Trinity College, Cambridge, entering the latter at age eighteen. Cambridge, especially Trinity College, has maintained an extraordinary reputation for mathematics, dating much farther back than Göttingen's. After all, it housed Newton, the father of calculus.

Cambridge also gave birth to a formulation of higher-dimensional space that predated even Riemann's. In 1843, mathematician Arthur Cayley, a fellow of Trinity College, published a paper entitled "Chapters in the Analytical Geometry of N Dimensions." It showed how one could define dimensionality as the number of coordinates needed to describe a space, and suggested ways to envision higher dimensions. The following year, the German mathematician Hermann Grassmann published an independent work on the same subject.

A related article by William Hamilton of another well-known Trinity College—in Dublin—also appeared in 1844. Entitled "On Quaternions," it introduced the concept of mapping aggregates of real and imaginary (the square roots of negative) numbers onto spaces of four dimensions. (Gauss had performed this feat already for two dimensions.) Hamilton suggested the idea of using three different representations for the square root of negative one, which he called i, j, and k. He assigned each its own special axis. Along with the conventional number line, this defined four perpendicular axes. He demonstrated that any number written in his notation (called quaternions) corresponded to a point in this four-dimensional framework.

Cayley later became a pioneer in the field of linear algebra, introducing the concept of matrices. Matrices are arrays of numbers, arranged in rows and columns. Just like ordinary values, matrices can be added, subtracted, and multiplied. They also can be used to transform one point to another in a space of arbitrary dimensions. Cayley demonstrated that quaternion theory could be fully expressed in terms of his matrix algebra.

As a student at Cambridge, Clifford investigated these subjects with great interest. Then when Riemann's work was finally published, it set his imagination ablaze. Clifford began to wonder if he could use non-Euclidean geometry and higher dimensions to encompass the physical as well as the mathematical world.

In 1870, he published "On the Space Theory of Matter," advancing the notion that everything in the world consists of bumps in space. In Clifford's view, not only could one describe fields (electric, magnetic, gravitational, and so on) through geometry, but also the particles that interact by means of those fields. The building blocks of materials, he argued, are non-Euclidean disturbances of a "flat" (noncurved) three-dimensional space. If we could somehow view them from the perspective of a higher dimension, they would be as noticeable as tire tracks in the desert. Because we are confined to space, however, these hyperspatial bulges present the illusion of solid bodies. Little was known at the time about the fundamental composition and structure of matter, so Clifford's explanations weren't very sophisticated. Nevertheless, like Riemann's thoughts on the subject, they foreshadowed in certain ways many of the notions of twentieth- and twenty-first-century theoretical physics, including relativity, Kaluza-Klein theory, and M-theory.

After leaving Cambridge, Clifford was appointed to the chair of Mathematics and Mechanics at University College, London. By all accounts he was an excellent teacher, spending considerable time with his students. All the while, he tinkered relentlessly with his spatial model of matter. A workaholic by nature, he would come home after a hard day's teaching and conduct his research throughout the night. In his spare moments he indulged in a passion for writing children's fairy tales. This ceaseless labor left him little time for sleep and took a great toll on his health. Consequently, like his hero Riemann, Clifford's life was cut short by pulmonary illness. He died two months before his thirty-fourth birthday.

A Plea for the Mathematician

Like Cayley and Clifford, the mathematician James J. Sylvester also spent his university years in a maze of courtyards and halls near the river Cam. A Londoner of Jewish heritage, he attended St. John's College, a friendly rival to Trinity. A feisty and independent lad, he never held back his views or conceded an argument due to social pressure. When about to graduate, Cambridge officials asked him, as a matter of routine, to take a Church of England religious oath. He refused, and never got his degree.

He began his long career in 1838, at the age of twenty-four. As a non-Anglican, he took one of the only positions open to him, the chair of natural philosophy at nonsectarian University College. This involved teaching introductory physics, including setting up labs and demonstrations. As someone interested in pure mathematics, spending his time with such practical details was only one step above washing the floors. He bolted to the University of Virginia, where a proper math professorship lay vacant for him.

Sylvester's sojourn to the States was notoriously short-lived. Within three months, he had to leave because of a disciplinary disaster in one of his courses. A student caught reading the newspaper in class responded by firing a volley of insults. Enraged, Sylvester poked him with a sword-cane, causing him to fall to the floor. Dramatically the student cried out, "I am killed," to shocked onlookers. Not seriously hurt, however, the student recovered far faster than Sylvester's feelings. After reprimanding the university authorities for their lax policies, Sylvester decided it was high time to leave. He fled to New York, where his brother Sylvester J. Sylvester was working on Broadway as a lottery agent. Exasperated by his time in the United States, he then sailed back to England.

By this point the doors of academia were barred to Sylvester, so he decided to work as an actuary for a London insurance firm. On the side, he gave private instruction in mathematics to selective students. Burnt by his experience in Virginia, he didn't want any rowdy ones, only "Florence Nightingales." He got what he wished for. Florence Nightingale herself, preparing for a career of reforming military hospitals, came to him for tutoring.

The next step along what Sylvester called "the world's slippery path"[6] was pursuing experience in law. At the venerable complex of wigged judges and frantic clerks known as Lincoln's Inn, he became acquainted with Cayley, who had also become a lawyer. They established a two-man club of Jekyll and Hyde researchers—transforming into idealistic pure mathematicians over lunch or during a walk, but then reverting back into hard-nosed legal authorities for the rest of the day. Sylvester and Cayley would be friends and collaborators for life, successfully pursuing matrix theory and higher-dimensional geometry. They jointly developed a "theory of invariants," determining which transformations leave certain quantities alone. For instance, changing a positive into a negative yields

James Joseph Sylvester, a respected nineteenth-
century British mathematician, was one
of the early developers and promoters
of hyperspace geometries.

the same squared values. This concept would later become an essential part of relativity.

Hankering for another chance to be a professor, in 1854
Sylvester obtained a position at the Royal Military College in Woolwich. Cayley took on a professorship at Cambridge in 1863. Together
they kept up with the exciting developments in non-Euclidean
geometry, marveling at the revolutionary theories of Gauss, Riemann, and Clifford.

In 1869, Sylvester became president of the Mathematical and
Physical Section of the British Association for the Advancement of
Science. He decided to use the occasion to make an impassioned
speech on behalf of what he saw as the cutting edge of the times: the
discovery of new mathematical realms, such as hyperspace, through
the application of novel methods and ideas. Published in two install-

ments of the first volume of the broadbased scientific journal *Nature,*
Sylvester's piece introduced a wide audience to the astonishing concept that there may be more dimensions than meet the eye.

Sylvester devoted a sizable part of his exposition to one of his
favorite subjects, refuting Kant's doctrine that space is a form of intuition. On the contrary, Sylvester argued, space is physically real and
can be explored through objective measurement. If mathematicians
reasonably surmise that hyperspace exists, it should similarly be
treated as something tangible. As Sylvester noted, "Mr. Clifford . . .
and myself . . . have all felt and given evidence of the practical utility
of handling space of four dimensions, as if it were conceivable
space."[7]

If hyperspace is real, then why can't humans directly experience
it? Sylvester addressed that question by means of an analogy he
attributed to Gauss,[8] involving a two-dimensional bookworm familiar
with nothing other than its own page: "As we can conceive beings
(like infinitely attenuated book-worms in an infinitely thin sheet of
paper) which possess only the notion of space of two dimensions, so
we can imagine beings capable of realising space of four or a greater
number of dimensions."[9]

In an extremely lengthy footnote to the *Nature* article, Sylvester
used this analogy to herald Clifford's spatial theory of matter (which
Sylvester learned about before it was published):

> Clifford has indulged in some remarkable speculations as to the
> possibility of our being able to infer, from certain unexplained
> phenomena of light and electromagnetism, the fact of our level
> space of three dimensions being in the act of undergoing in space
> of four dimensions (space as inconceivable to us as our space to the
> supposititious bookworm) a distortion analogous to the rumpling
> of the page.[10]

After Sylvester's talk appeared, *Nature* became home to a running dialogue about the possibilities of higher dimensions. Letters
to the editor served as position papers for the debate. Sylvester's
bookworm argument became the subject of various nuances of
interpretation. This heightened interest led Clifford in 1873 to
translate and publish Riemann's speech for the benefit of *Nature*'s
readers. Once the general scientific audience became familiar with

hyperspace, it would not be long before it entered the popular imagination as well.

Parlor Tricks

By the late nineteenth century, the winds of scientific change had been blowing for hundreds of years from the university centers of Europe, but much of the public still clung to the old ideas. From the simple villages of the Harz lands northeast of Göttingen to the opulent terraces of Mayfair in London, those of a superstitious bent clutched tightly to tales of ghosts, fairies, and witches. As science enveloped more and more territory, however, it became unclear where on Earth the legendary figures could reside.

Consequently, once the concept of hyperspace emerged there were many who pondered that this realm was full of spirits. If science encompassed the visible three-dimensional universe, then perhaps the mysterious angels and demons of yore could find ample home in a place beyond space. (One writer, A. T. Schofield, would in 1888 even suggest that God lives in the fourth dimension.)

The public perception of a link between higher dimensions and mysticism was cemented in 1877 when the German physicist Johann Zöllner defended American medium Henry Slade against accusations of fraud during a sensational trial in London. Slade, who drew attention to himself by conducting séances with prominent Londoners, was charged with "using subtle crafts and devices, by palmistry and otherwise," to deceive his followers. Zöllner called for a thorough scientific inquiry into Slade's abilities.

Under the watchful eye of Zöllner and several other witnesses, Slade performed a number of seemingly impossible tricks. He linked solid wooden rings together, transported objects out of sealed containers, removed the knot from a tied rope whose ends were attached together, and produced written messages on paper trapped between solid slabs of slate. Zöllner was mesmerized by these feats, concluding that the only possible explanation was that Slade had found a way to move things through a dimension other than length, width, and height.

As if announcing a revolutionary scientific discovery, Zöllner enthusiastically reported his conclusions to the public. In his mind, he had discovered the portal to a whole new world. Even after skeptics pointed out that any good magician could replicate Slade's tricks, Zöllner wrote assuredly in several scholarly works that higher dimensions were real. Consequently, as the American architect Claude Bragdon wrote, "Zöllner's name became a word of scorn, and the fourth dimension a synonym for what is fatuous and false."[11]

From that point on those interested in the scientific possibility of a realm beyond the ordinary dimensions often felt compelled to emphasize that their views had nothing to do with mysticism. Yet for every serious scientist or mathematician laying down his arguments, dozens of occultists would invoke the fourth dimension to justify their beliefs. The expression "another dimension" became tantamount to the world of the spirits, a connotation that remains today.

The rise of the Theosophy movement, founded by Helena P. Blavatsky in 1875, generated even more interest in the mystical aspects of higher dimensions. Theosophy is an occult system of beliefs that draws from many sources, including Kabbalistic texts, Vedic doctrines, and transcendental Greek writings. Blavatsky believed she could use otherworldly knowledge to understand the properties of matter and other aspects of science. She maintained, for instance, that materials could pass through one another—as in Slade's feats—by transforming their essential properties.

Inspired by Zöllner, many Theosophists embraced the fourth dimension as a way of explaining both spirit and substance. Blavatsky herself was skeptical of this view. In *The Secret Doctrine,* her best-known work, she refuted the idea of a fourth spatial dimension and argued instead for new ways of understanding the characteristics of matter. As she explained:

> [W]hen some bold thinkers have been thirsting for a fourth dimension to explain the passage of matter through matter, and the production of knots upon an endless cord, what they were really in want of, was a *sixth characteristic of matter.* The three dimensions belong really but to one attribute or characteristic of matter—extension; and popular common sense justly rebels against the idea that under any condition of things there can be more than

three of such dimensions as length, breadth, and thickness. These terms, and the term "dimension" itself, all belong to one plane of thought, to one stage of evolution, to one characteristic of matter.[12]

Despite Blavatsky's views, many Theosophists maintained interest in the fourth dimension, equating it to the concept of the "astral plane." According to influential Theosophist C. W. Leadbeater, who wrote a book on the subject, the astral plane is the nonphysical domain in which clairvoyance and other occult phenomena supposedly takes place. As Leadbeater emphasized, "Short of really gaining the sight of the other planes, there is no method by which so clear a conception of astral life can be obtained as by the realization of the Fourth Dimension."[13]

Other followers of Theosophy who preached the mystical significance of the fourth dimension included Bragdon and the Russian occultist P. D. (Peter) Ouspensky. Bragdon wrote a number of popular books on the subject and incorporated hyperspace-like elements into his architectural designs. He also participated in the translation of Ouspensky's *Tertium Organum,* a book that purported to resolve the world's major enigmas.

The Society of Psychical Research, founded in 1882 as a group devoted to the scientific study of paranormal experiences, has maintained similar interest in higher dimensions. Founding members of the society included the philosopher-psychologist William James and the physicist William Crookes, a strong supporter of Zöllner. The activities of the society were famously satirized in Oscar Wilde's "The Canterville Ghost," which mentions a ghost escaping from a room by means of a fourth spatial dimension.

Slicing up the Hypercube

While the Zöllner controversy was raging, Sylvester was embarking on his second mathematical career. After being forced to retire from Woolwich in 1870, Sylvester had thought that his academic life was over. He retreated to his London home and spent much of his time reading and writing poetry. He was especially proud of a book on versification that he wrote, *The Laws of Verse.* Then, in 1876, word came from the newly founded Johns Hopkins University in Balti-more that they wanted him to fill a mathematical professorship. Even though his prior experience in the United States had left a bad taste in his mouth, he cheerfully accepted the call—at the age of sixty-two—to take on a prestigious new role.

Once Sylvester settled in at Hopkins he was like a kid in a candy store. He was thrilled by the possibility of shaping the careers of graduate students, a pleasure he didn't have in England. They were so friendly and eager, he didn't have to wield a sword-cane even once. He also relished the thought of promoting new advances in mathematics. To that end, he founded the *American Journal of Mathematics,* a publication that in its early years served in part as a showcase for the work of his students and associates.

One of Sylvester's prize students, W. Irving Stringham, had a Clifford-like knack for spatial visualization. Applying this talent to his adviser's pet subject, he constructed a menagerie of representations of hyperspace. These depicted what the six regular four-dimensional objects (in analogy to the Platonic solids) would look like if sliced by the three-dimensional space of our awareness. Stringham used Euler's theorem, a relationship between the number of faces, edges, and corners of geometric objects, to help construct these. In 1880, he published his artfully drawn images in the *American Journal of Mathematics.*

The hypercube, one of Stringham's studies, has subsequently become one of the most commonly depicted four-dimensional objects. It forms the natural successor to the point, line segment, square, and cube. In a plane geometry, a point moved in any direc-tion traces out a line segment. A line segment, transported by its own length parallel to itself, constitutes a square. Moving a square in sim-ilar manner produces a cube. Therefore, by simple extension, to pic-ture a hypercube one might imagine transporting a cube parallel to itself through the fourth dimension. Because one obviously cannot do this on paper, one is left with the less satisfying option of sketch-ing two images of a cube, then drawing line segments connecting the corners of one with the matching corners of the other. Only with the recent advent of animated computer graphics have researchers (such as Brown's Thomas Banchoff) been able to improve substan-tially upon this approach.

Line Segment   **Square**   **Cube**

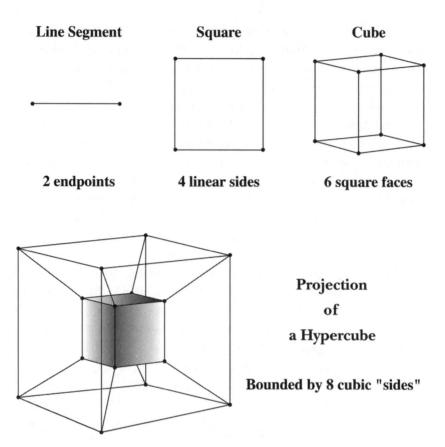

2 endpoints   **4 linear sides**   **6 square faces**

Projection

of

a Hypercube

Bounded by 8 cubic "sides"

One can arrange four line segments to make a square, and six squares to form a cube. By extrapolation, one might envision eight cubes forming a hypercube.

After receiving his Hopkins degree, Stringham took on a position in Leipzig, Germany, working with the noted mathematician Felix Klein (no relation to Oskar Klein, coauthor of Kaluza-Klein theory—and neither Felix nor Oskar was the inspiration for Neil Simon's famous play). Klein, who made substantial contributions to non-Euclidean geometry and developed the Klein bottle, was a keen believer in physical models of abstract mathematics. Stringham's designs and Klein's advocacy encouraged the mathematician Victor Schlegel to construct actual three-dimensional representations of hypersolids. These "hyperdimensional toys" were first displayed in 1884 at a physician's convention, and were later made available for purchase. Göttingen's math department acquired a number of these, which are prominently showcased even today.

In 1883 Sylvester received the exciting news that Oxford University was interested in appointing him professor. Although he was happy at Hopkins, he missed England, and once again decided to make a transatlantic voyage. He generously aided Hopkins officials in finding a suitable replacement. He enlisted Stringham to inquire about Klein's eligibility for the job, and also invited Cayley to apply. Ultimately the position went to Simon Newcomb, a Canadian-born astronomer, who inherited the *American Journal of Mathematics* as well as Sylvester's interest in hyperspace.

After happy years at Oxford, Sylvester died in 1897 at the age of eighty-three. He worked on his mathematical projects virtually until the end. Cayley's death preceded that of his friend by two years. Perhaps no other British mathematicians did more to further the exploration of higher-dimensional geometries. Cayley's quiet discoveries, matched by Sylvester's boundless enthusiasm and bold rhetoric, helped inspire a generation of thinkers to make hyperspace their home.

Tesseract Construction Kits

From Jules Verne to J. K. Rowling, the literature of the fantastic has had a variety of settings. Some of these are real places—the moon, for instance, or the center of the earth. Others are wholly imaginary, such as dragon's caves or wizard's schools. Stories about higher dimensions fall into a third category, concerned with extensions of physical reality that may or may not truly exist.

In the late nineteenth century, three visionary writers, all from London, pioneered the genre of hyperspace tales. The first, C. Howard Hinton, was better known by his contemporaries, including notables such as William James, than by current readers. Yet his influence lives on through those inspired by his writings. On the other hand, who today is unfamiliar with H. G. Wells, author of *The Time Machine, The Invisible Man, The War of the Worlds,* and other famous stories? Along with Verne, Wells practically invented science fiction. Somewhere in between stands Edwin Abbott Abbott, well known for a single novel, *Flatland,* but little known for his other writings.

Charles Howard Hinton, born in 1853, was the son of the prominent surgeon and libertine writer James Hinton. Because of his father's outlook, the younger Hinton was exposed at an early age to quite liberal attitudes toward sexuality. His wife, Mary, the daughter of the notable late mathematician George Boole, came from a similarly permissive household. With such a background, Hinton had diverse choices in life, including becoming a mathematics professor, a writer, or a sexual outlaw. He checked off all of the above, guaranteeing an exciting but notorious lifestyle.

His teaching career began with the post of headmaster at Uppingham School. He wasted no time starting his second occupation. Beginning in 1880 with the article "What Is the Fourth Dimension?" he published numerous essays and stories in the fields of mathematics and science. Although he wrote about many things, his focus was on ways to envision hyperspace.

Memory and forgetfulness played a great role in Hinton's personal philosophy. He was convinced that people could erase their own thought patterns and reorient their brains to encompass new modes of perception. To that end, he set out to develop a system by which anyone could learn how to "see" the fourth spatial dimension. A believer in Kant's idea that space is a form of intuition, he asserted that for the trained mind, hyperspace could become just as intuitive.

Hinton's technique involved assigning names and colors to the building blocks of four-dimensional objects, to help the mind remember their configurations. He coined the name *tesseract* to denote the hypercube, and devised the terms *ana* and *kata* to define the two ways one can move in the fourth dimension (analogous to up and down). Making use of his color scheme, he then detailed a step-by-step way by which one can picture a tesseract moving through ordinary space.

Hinton realized that no one could fathom a four-dimensional object all at once. For this reason, his method relied on the gradual progression of a hypersolid over time, as if a camera slowly recorded its motion. Such a technique is used in contemporary films trying to depict the enormity of a spaceship or other vast structure (the

Titanic, for instance). Instead of showing the full view of the craft, the director might choose to pan slowly from one end to the other, giving the audience a better chance of appreciating its bulk. Similarly, Hinton believed four-dimensional bodies could be appreciated only by imagining their evolving form as they pass through our space. As he emphasized, "All attempts to visualize a fourth dimension are futile. It must be connected with a time experience in three space."[14]

Despite his admonition, one of Hinton's images of a tesseract has become virtually an icon. Hinton noted that one could cut the surface of a cube and flatten it out in such a way that it resembles a cross. Four of the faces of the cube comprise its vertical staff, and the other two form its horizontal bar. Similarly, one can envision cutting the outside of a tesseract and unfolding it into a three-dimensional cross. In that case, four cubes stack up vertically and the other four become two perpendicular horizontal bars. Surrealist Salvador Dali would incorporate this striking image into his crucifixion painting *Christus Hypercubus.*

Hinton had an intriguing theory about the physical inaccessibility of the fourth dimension. He proposed that like the near-flatness of coins or cardboard cutouts, we only jut out a little bit in that direction. The bulk of an individual is three-dimensional, with the four-dimensional part being imperceptibly thin. By proposing the minuteness of a higher dimension as the reason for its inability to be observed, Hinton's theory curiously anticipated the Kaluza-Klein approach.

Mary Hinton was very supportive of her husband's intellectual pursuits. Regarding his personal life, she had to be more than just encouraging; she needed nerves of steel. While at Uppingham, he brought around a woman named Maude, whom everyone assumed was his sister. It turned out that he was dually married, to Maude as well as Mary. His bigamy extended to fathering twins with his second wife. When the authorities found out, he was jailed for three days. After his release, he and Mary fled England for Japan, then later for the United States. All the while, Mary, perhaps because of her liberal upbringing, steadfastly stood by her man.

While in the United States, Hinton managed to obtain an instructorship in Princeton's math department. He spent most of his

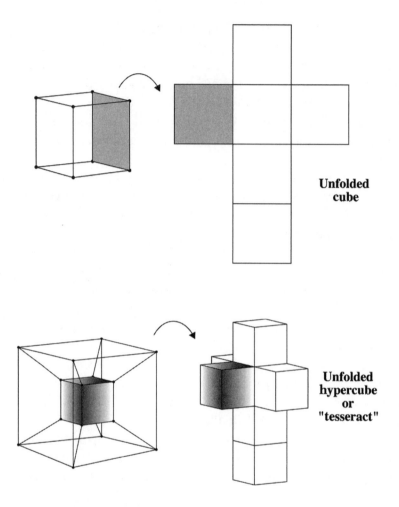

A cube, opened up and flattened out in a plane, can look something like a cross. Similarly, a hypercube, unfolded into three-dimensional space, can appear something like a three-dimensional cross. (Some of the cubes forming the cross are occluded from view because of the perspective.) This motif was explored by Salvador Dali in his painting Christus Hypercubus.

time developing a "baseball gun" for Princeton's team. With its capacity to fire balls at up to 70 miles per hour, the gun helped relieve pitchers during practice. It even had a special attachment for curve balls. Though the coaches were no doubt appreciative, Hinton was fired.

Subsequently, after another teaching stint in Minnesota, Hinton was assisted by Simon Newcomb in obtaining a position at the Naval Observatory in Washington, D.C. Because of his fascination with higher dimensions, Newcomb was interested in Hinton's work and wanted to see the exiled thinker regain his footing. Hinton lived in Washington until his untimely death in 1907.

The manner of Hinton's death was quite unusual, a bizarre capstone to an extraordinary life. While attending a banquet for the Society of Philanthropic Inquiry, the master of ceremonies proposed a toast to female philosophers. Hinton picked up his glass to honor the request and then simply dropped dead. Indicative, perhaps, of the priorities of the times, his obituary emphasized his invention of the baseball gun as much as it did his writing.

Unrolling Flatland

Edwin Abbott Abbott was another headmaster-turned-writer fascinated by higher dimensions. The similarities to Hinton end there. In lifestyle at least, Abbott was as devout as Hinton was irreverent, viewing faith and hard work as a critical part of being. For him "slackness, sloth and deceit were almost unpardonable sins."[15] That said, Abbott's religion was not the stodgy variety, but rather blossomed as a tolerance toward people of various backgrounds and a supreme love of education.

Born in 1838, Abbott attended St. John's College at Cambridge and became an ordained Anglican priest. At the unusually youthful age of twenty-six, he was recruited to lead the City of London School. Located in the heart of the city, the school was a haven for Londoners of many different classes and creeds. Abbott relished this diversity, and turned down opportunities to administer more elite schools. Rather, he shaped the City of London School into arguably the most progressive, innovative, and scientifically advanced educational center for boys at the time. When the crowded school outgrew its quarters, Abbott argued passionately for a more spacious structure with state-of-the-art laboratories, comfortable lecture halls, and ample playgrounds. His interest in scholastic reform echoed Dickens and other nineteenth-century progressive thinkers.

The new building Abbott built was a beauty. Situated right on the banks of the Thames, it looked more like a Renaissance palace than a day school. Its gleaming white stone and brick facade was fronted by statues of notable Englishmen, including Francis Bacon, William Shakespeare, John Milton, and Isaac Newton. The statues advertised well the pursuits of the headmaster inside. Abbott was a respected authority on Baconian philosophy, Shakespearean grammar, biblical interpretation, and physical science.

The completion of Abbott's dream edifice meant that he had greater time for an even more satisfying kind of construction: the joy of putting down his thoughts on paper. Invoking Old Testament imagery about Joshua replacing Moses once the Promised Land was entered, he briefly considered resigning to become a full-time writer and scholar. Nonetheless, he remained steadfast at his post for seven more years. At the same time he produced some of his most creative works, including a fine biography of Bacon and, most famously, *Flatland,* his wonderful, multidimensional scientific romance.

In his Bacon biography, Abbott praised the English philosopher's independence of thought and described many of the obstacles he faced in combating prejudice against scientific progress. Abbott showed how Bacon particularly fought against Aristotelian rigidity. The protagonist of *Flatland* wages a similar battle, proving to his own society's "Aristotelians" that his world has extra unseen dimensions. The book is dedicated to the "inhabitants of space" so that they "may aspire yet higher and higher to the secrets of four, five or even six dimensions, thereby contributing to the enlargement of the imagination."[16]

Abbott wrote *Flatland* from its leading character's point of view, and framed it as if it were a miraculously discovered diary. In the first edition, published in 1884, Abbott's name is nowhere to be found. Instead it is signed the name of the protagonist, A Square. As pointed out by the mathematician Rudy Rucker, this may have been a clever play on the fact that Abbott's middle and last names were identical, like squaring a quantity.[17] At any rate, A Square bears the geometric appearance of his name and lives in a two-dimensional world of many other such shapes.

Despite the title, *Flatland* is a story with considerable depth that works on many levels. Abbott described a supremely hierarchical society, confined to a plane, in which a character's number of sides strictly

determines his fate (the more, the better). Triangles are the hoi pol-
loi, with the isosceles variety—the common workers and soldiers—
even lowlier than the equilateral sort. As the professional classes,
Squares—such as the story's hero—and Pentagons command somewhat greater respect. As noblemen, Hexagons outrank Pentagons,
and so forth. Heading the society are the Circles, serving as priests. On
the very lowest rung are the women, who as basic line segments (or,
alternatively, exceedingly thin triangles) are in poor shape indeed.

It's not hard to see what culture Abbott is mocking, since he
wrote during the corset-tight ethos of the Victorian era. By not only
describing a ridiculously stratified society, but by also setting it in an
isolated sliver of space—an island kingdom, like Britain—he parodied the inability of his countrymen to look beyond themselves.
Even after A Square discovers the truth and tries to reveal it to them,
the Flatlanders refuse to believe that higher dimensions exist. Ironically, thanks to Cayley, Clifford, Sylvester, Hinton, and others, the
Victorians were the first generation of Britons to contemplate seriously the meaning of higher dimensions. In that respect, they were
less closed-minded than Abbott's satire would suggest.

A Square's realization that life exists beyond his plane is presented as a quasi-religious experience. The revelation takes place
during the dawning of the third millennium (according to Flatland's
calendar), a time of great anticipation. While conversing with his
wife, A Square suddenly "became conscious of a Presence in the
room." It was a great Circle "that seemed to change its size in a manner impossible for a Circle or for any regular Figure of which [he]
had experience."[18]

The Circle turns out to be a Sphere that has emerged through
the plane of Flatland. As the Sphere rises, it presents itself to the Flatlanders as circles of increasing diameter, like slices of a round loaf of
bread. Having no eye that can look beyond the plane in which he
lives, A Square cannot see the whole, at first, just a succession of
parts. This is similar to Hinton's method for envisioning a tesseract.

After announcing that he has come to convey the "Gospel of the
Three Dimensions," the Sphere appoints A Square his apostle. A
Square is incredulous, until the Sphere physically lifts him off his
surface and into Spaceland. While flying over his country, tugged
like Wendy by Peter Pan, he views a panorama of houses, trees, and

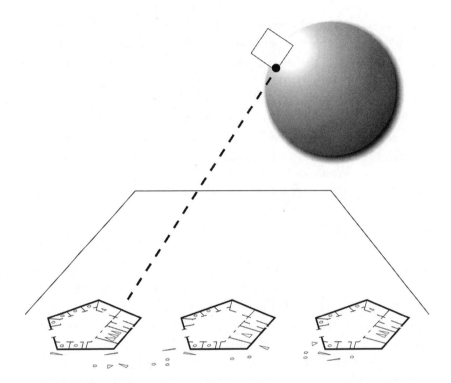

A Square observes the panorama of two-dimensional Flatland as Sphere lifts him high above it along the third dimension. From his new vantage point, he can simultaneously see the inside and outside of all people, places, and things in his community.

people. Unlike Wendy, however, he can see the insides of everything (and everyone) as well as their outsides. From the vantage of space, nothing in the plane escapes his eye, from the furniture of his apartment to the intestines of his wife. In short order, the four-sided hero is an enthusiastic convert to three-dimensional belief, and hopes to spread the good news to his fellow countrymen. Predictably, they think he's insane and lock him up. He spends his years of incarceration bemoaning his inability to convey the truth about space. As he laments at the novel's conclusion:

> Hence I am absolutely destitute of converts, and, for aught that I can see, the millennial Revelation has been made to me for nothing. . . . Yet I exist in the hope that these memoirs, in some manner,

I know not how, may find their way to the minds of humanity in Some Dimension, and may stir up a race of rebels who shall refuse to be confined to limited Dimensionality."[19]

Indeed, A Square's work, as channeled through Abbott, was hardly for naught. Generations of delighted readers have used *Flatland* as a launching pad for expeditions into fantastic mathematical conceptions. And it has spun off many clever sequels, each emphasizing a different aspect of dimensionality, including Dionys Burger's *Sphereland*, A. K. Dewdney's *The Planiverse*, and most recently, Ian Stewart's *Flatterland*.

Highway to the Future

Throughout most of the nineteenth century, as higher *spatial* dimensions became an integral part of mathematics, few writers contemplated the possibility of *time* as the fourth dimension. The simple proposal of d'Alembert and Lagrange was drowned out by a chorus of interest in the bizarre worlds of hyperspace and non-Euclidean geometry. Warped spaces, parallel lines coming together, and secret enclaves of the spirits presented a much more entertaining spectacle than merely pointing to the clock.

By the late nineteenth century, however, the pendulum started to swing back. Hinton's conception of a tesseract evolving through space and Abbott's image of a sphere displaying increasing and decreasing circles as it moves through a plane suggested a strong temporal aspect to four-dimensionality. It wouldn't be long before thinkers would revisit the notion of linking extension with duration.

In 1885, a writer who signed his name "S" published an insightful piece in *Nature* entitled "Four-Dimensional Space." (The physicist James Beichler has postulated that "S" was indeed Sylvester, who sometimes signed his name that way.[20]) The article foreshadowed the relativistic vision of Einstein and Minkowski by proposing an amalgamation called "time-space."

We must . . . conceive that there is a new three-dimensional space for each successive instant of time; and, by picturing to ourselves the aggregate formed by the successive positions in time-space of a given solid during a given time, we shall get the idea of a four-dimensional solid, which we may call a sur-solid. . . . Let any man picture to himself the aggregate of his own bodily forms from birth to the present time, and he will have a clear idea of a sur-solid in time-space.[21]

Around the time this article was published, Herbert George Wells, a student at the Royal College of Science (now part of Imperial College) in London, likely first learned about the fourth dimension from his fellow classmates. Born in Bromley (south London) in 1866 as the son of a tradesman, Wells began college at the age of eighteen. His first teacher, the esteemed biologist Thomas Huxley, introduced him to the revolutionary new theory of evolution. Wells became so excited by science that he, along with some friends, decided to found and edit a paper called the *Science School Journal.* In April 1887, one of his fellow students, E. A. Hamilton Gordon, published an article in the journal entitled "Fourth Dimension." Wells became fascinated by the topic and, very shortly thereafter, contributed a story about it, "The Chronic Argonauts," to the same journal.[22]

Wells became a prolific writer and would explore the concept of the fourth dimension in many of his other short stories and novellas, including *The Invisible Man,* "The Wonderful Visit," "The Remarkable Case of Davidson's Eyes," and "The Plattner Story." In each of these tales, he used the fourth dimension as a plot device to justify unearthly phenomena, such as human bodies transparent to light, universes parallel to our own, eyes able to see objects thousands of miles away, inverted anatomies, and more. For example, in "The Plattner Story," a tale with a Möbius-like twist, he wrote, "The curious inversion of Plattner's right and left sides is proof that he has moved out of our space into what is called the Fourth Dimension."[23]

Undoubtedly, Wells's most famous treatment of dimensionality was in *The Time Machine,* his 1895 novella based on "The Chronic Argonauts." It tells the tale of an inventor who perfects a device that can sail through time. He uses the machine to travel hundreds of thousands of years into the future, meeting the strangely evolved descendants of the human race. Returning to the Victorian era, he reports his findings to his incredulous friends, before vanishing with his machine once more, never to be seen again.

The way the Time Traveller (as he is called) explains the ma-

chine's workings to his friends bears striking resemblance to the
theory expressed by "S." Wells argues that time travel is possible
because time and space are part of the same entity. Compare the idea
of sur-solids and time-space to this passage from the story: "Here is a
portrait of a man at eight years old, another at fifteen, another at
twenty-three, and so on. All are evidently sections, as it were, Three-
Dimensional representations of his Four-Dimensional being, which is
a fixed and unalterable thing."[24]

It is not clear if Wells had read the article by "S," or if he was
merely familiar with the general notion of time as the fourth dimension. Interestingly, in *The Time Machine* Wells mentioned a speech by
Newcomb with a somewhat different approach that also appeared in
Nature: "Some philosophical people have been asking why *three*
dimensions particularly—why not another direction at right angles
to the other three?—and have even tried to construct a Four-
Dimensional geometry. Professor Simon Newcomb was expounding
this to the New York Mathematical Society only a month or so ago."[25]

The allusion is to "Modern Mathematical Thought," an address
Newcomb delivered before the New York Mathematical Society
(shortly before it became the American Mathematical Society) in
December 1893. In the talk he emphasized the difference between
mathematicians' and physicists' concept of the fourth dimension. As
Newcomb stated:

> As the boy, at a certain stage in his studies, passes from two to three
> dimensions, so may the mathematician pass from three to four
> dimensions with equal facility. . . . The mathematician, if placed
> inside a sphere in four-dimensional space, would simply step over it
> as easily as we should over a circle drawn on the floor. Add a fourth
> dimension to space, and there is room for an indefinite number of
> universes, all alongside of each other, as there is for an indefinite
> number of sheets of paper when we pile them upon each other.
>
> From the point of view of physical science, the question
> whether the actuality of a fourth dimension can be considered
> admissible is an interesting one. All we can say is that, so far as
> observation goes, all legitimate conclusions seem to be against it.
> No induction of physical science is more universal or complete
> than that three conditions fix the position of a point. The phe-
> nomena of light shows that no vibrations go outside of three-
> dimensional space, even in the luminous ether. If there is another

universe, or a great number of other universes, outside of our own, we can only say that we have no evidence of their exerting any action upon our own.[26]

Newcomb's speech anticipated some of the central questions of twentieth- and twenty-first-century theoretical physics. What physical meaning might science derive from the existence of higher dimensions in mathematics? Assuming time is the fourth dimension, as Einstein and Minkowski would later show, might there be unseen spatial dimensions as well? If so, what would be their physical consequences? If there are no effects, then how might we explain their inability to be observed?

These questions have come to the forefront during the scientific search for a unified theory of all natural forces. The quest for a Theory of Everything, elegantly explaining all aspects of the diversity and unity of nature, has become one of the most compelling goals of science. Many physicists believe this can be achieved only through an understanding of higher dimensions beyond space and time.

CHAPTER 3

The Physicist's Stone

Uniting Electricity, Magnetism, and Light

The non-mathematician is seized by a mysterious shuddering when he hears of "four-dimensional" things, by a feeling not unlike that awakened by thoughts of the occult. And yet there is no more commonplace statement than that the world in which we live is a four-dimensional spacetime continuum.

—ALBERT EINSTEIN, *Relativity*

Nature's Unity

Nature is a study of vivid contrasts and subtle connections. The brilliant theatrics of lightning blazing across the sky and the soft glow of luminescent fish gliding through the deep. The shrieking gales that set in when a storm front hits and the gentle ripples that appear when a pebble drops into a still pond. The startling clamor of an avalanche releasing half a mountain into an innocent valley and the quiet procession of the moon, marking the somnolent hours with its passage. Our senses speak to the diversity of these phenomena, but our minds note common physical origins.

Over the years, science has learned that a natural force can wear many guises. The same electric forces that mercilessly fell a giant tree might later innocently stick together tiny bits of paper derived from its pulp. The same gravitational powers that hurriedly pull billions of tons of water into a tidal basin might later offer children the quiet joy of skipping stones along the basin's surface. Mighty or mellow, the faces are different but the actors the same.

Since the time of Newton, scientists have looked for a comprehensive way of explaining the unity and diversity of the world. Newton's brilliant deduction that the same force of gravity attracting objects down to Earth also steers the planets in their orbits around the sun led to a search for underlying descriptions of all natural forces. With this goal in mind, beginning in the eighteenth century, researchers began to explore the properties of electricity and magnetism as well.

In 1785, the French physicist Charles-Augustin Coulomb successfully used a device called a torsion balance to measure the electric forces between charges and the magnetic force between poles. He discovered that, like gravity, each force obeyed an inverse square law. In other words, these forces tapered off as the square of the mutual distance between either the charges or poles under investigation. For instance, if you take two charged spheres and double their mutual distance, their electrical force goes down fourfold. If you move magnetic poles apart in the same way, you get similar reduction in magnetic force.

In the decades that followed, other researchers such as André-Marie Ampère, Hans Christian Oersted, Wilhelm Weber, Joseph Henry, and Michael Faraday experimented with the properties of fluctuating electrical currents and changing magnetic forces. Their work led to speculations about deep connections between electricity and magnetism. By the mid-nineteenth century, the wealth of information gathered about electrical and magnetic interactions called out for a second Newton to elucidate the underlying principles of these forces.

Swift Currents

Growing up on a country estate, bordered by a rugged river, nestled in the untamed moors of Scotland, James Clerk Maxwell couldn't help but be keenly aware of the wild splendor of nature's drama. As the lonely boy walked along the rocky banks of the turbulent Water of Urr, he listened intently as its gurgling voice was met by the moans of sweeping winds. He marveled at the guttural music of the frogs as they celebrated their mysterious transformations from mere tadpoles. The dialogue of the elements interested him much more than the chatter of people. His brilliant mind strove for ways to decipher this language and unlock the hidden patterns of the world. His mother taught him that all beauty emanated from God. He deeply believed this, but wanted to understand the mechanisms underlying this grandeur.

In 1839, when Maxwell was only eight years old, his mother died. He withdrew even further into his own thoughts. After being tutored for a few years, he was sent off to the Edinburgh Academy. There he was mocked by his fellow students, who gave him the cutting nickname Dafty. They couldn't understand why he preferred drawing complex diagrams and making mechanical models more than playing ordinary childhood games.

After only six years of formal schooling, Maxwell entered Edinburgh University, where his professors included William Hamilton, the founder of quaternion theory. There his youthful interests blossomed into a love of experimental science and a knack for higher mathematics. Soon he was off to Cambridge to complete his studies.

As Maxwell became immersed in the grind of university life, achieving splendid success with his hard work, he often pined for nature's freedom. Though he had settled into a routine as staid as the river Cam, the raging waters of his youth coursed through his veins. Like Thoreau and Whitman, he yearned for deep connection with powerful natural forces. "Oh that men indeed were wiser," he wrote during a particularly dreary day, "and would raise their purblind eyes to the opening mysteries."[1]

To help unravel such enigmas, in 1855 Maxwell turned his considerable analytical powers to the workings of electricity and magnetism. At that time there were two major approaches toward the subject. One method, based on the Newtonian description of forces as "actions at a distance," envisioned point charges (or poles) tugging one another by means of imaginary cords. Maxwell found this too abstract and paid it little heed.

The other approach, pioneered by Faraday, pictured the same agents influencing each other through weblike connections that permeate all of space. This is the basis of the field concept, although Faraday, unfamiliar with mathematics, didn't express it that way. He came upon the idea by mapping the distribution of iron filings and other suitable materials near systems of charges, currents, and magnetic dipoles (the north and south poles of a bar magnet, for instance). Faraday summarized his results in the three-volume treatise *Experimental Researches in Electricity.*

Maxwell was galvanized by Faraday's book, attracted in particular to its tangible physicality. Setting out to develop a mathematical language to describe Faraday's experimental discoveries, Maxwell came upon the analogy of fluid flow. He realized that the field lines emanating from a positive charge (or north magnetic pole) spread out like a fountain of water. Those lines approaching a negative charge (or south magnetic pole), on the other hand, came together like water flowing into a drain. Thus, referring to the positive charges as "sources" and the negative charges as "sinks," he employed the equations of hydrodynamics (water flow) to map out the structure of electric fields. He used similar techniques to describe magnetic fields as well. In doing so, he demonstrated that the field concept was superior to considering forces simply as actions at a distance. Maxwell would later celebrate his triumph by penning these poetic lines:

> Thy reign, O Force! is over. Now no more
> Heed we thine action;
> Repulsion leaves us where we were before,
> So does attraction.[2]

In 1856, Maxwell left Cambridge and took on positions at Marischal College in Aberdeen, then King's College in London. There he discovered astonishing connections between electricity, magnetism, and optics. Four basic equations, he found, could express how electric and magnetic fields influence each other, as well as how they depend on the charges and currents that produce them. Simply stated, charges produce electric fields, and moving charges, or currents, create magnetic fields. Changing magnetic fields generate electric fields, and vice-versa.

By solving these relationships, now known as the Maxwell equations, he deduced the properties of electromagnetic waves: oscillations of electric and magnetic fields. These propagate through space like cars edging up one by one in a traffic jam. As one moves, it nudges the next, and so on. Calculating the velocity of these waves, he found it to be identical to the speed of light. Consequently, he arrived at the revolutionary conclusion that light is an electromagnetic wave.

In Maxwell's time, scientists believed that all waves must move through substances. Ocean waves travel through the water, and earthquakes through the land. In that case, what did electromagnetic waves move through? Maxwell and others speculated that space was filled with an invisible substance, called the aether. Light, as an electromagnetic wave, thus consists of vibrations in the aether.

In his later years, Maxwell returned to Cambridge, where he helped found the prestigious Cavendish physical laboratory. There he noted his colleagues' growing fascination with theories of higher dimensions. He followed with keen interest Clifford's attempts to express all material interactions through geometry. He also took note of the higher-dimensional work of Arthur Cayley and Felix Klein. Maxwell was even curious about the bizarre theories of his friend Peter Tait, who in a book with Balfour Stewart, *The Unseen Universe*, postulated that human souls were vortices (twists) in the aether. Like the knots tied by Henry Slade, these could only be untwisted in the fourth dimension. In one of the last poems he wrote, Maxwell gently poked fun at these theories:

> My soul is an entangled knot,
> Upon a liquid vortex wrought
> By Intellect, in the Unseen residing,
> And thine cloth like a convict sit,
> With marlinspike untwisting it,
> Only to find its knottiness abiding;
> Since all the tools for its untying
> In four-dimensioned space are lying
> Wherein thy fancy intersperses
> Long avenues of universes,
> While Klein and Clifford fill the void
> With one finite, unbounded homaloid,
> And think the Infinite is now at last destroyed.[3]

For his own taste, Maxwell preferred much more down-to-earth theories of nature involving realistic physical analogies. He liked to

think of his own electromagnetic model as describing the disturbances of an actual fluid, which is how he viewed the aether, rather
than something abstract. It is with some irony then that within three
decades after Maxwell's death (in 1879) scientists would banish the
aether and reconstitute his theory within a four-dimensional geometric context.

Clash of the Theories

In almost every way, Maxwell's theory of electromagnetism was a
resounding success. It brought unity to two out of three of the then-
known natural interactions. With electricity and magnetism united
as the electromagnetic force, only gravity was left out. Furthermore,
because the field description seemed much more flexible than the
idea of action at a distance, its introduction raised hopes that gravity
could be included under its umbrella as well.

A boon for the experimentalist, Maxwell's conception offered a
host of predictions about the nature of light. As an electromagnetic
wave, it determined, light should come in a wide range of frequencies, including many beyond the visible range. Sure enough, in 1888,
the German physicist Heinrich Hertz produced the first radio waves,
bearing much lower frequencies than optical light. Hertz generated
these through oscillating currents, the prototype of modern television and radio broadcasts. Infrared, ultraviolet, X-rays, and other
invisible forms of radiation were soon to be discovered. Maxwell's
equations also explained how radiation can have pressure, a result
confirmed in 1901.

There were several major difficulties with Maxwell's theory,
however—not with the equations themselves, but rather with how they
were interpreted. One was a conceptual point. Maxwell's assumption
that electromagnetic waves moved through aether was not borne out
by experiment. As hard as they tried, scientists could not detect such a
substance. Still, because it was difficult for them to imagine waves traveling through nothing at all, they held out hope that somehow the
aether would be found.

A related difficulty had to do with the relationship between Maxwell's equations and Newtonian dynamics. While traveling through

a given material, or even a complete vacuum, Maxwell's theory dictates that the speed of light must stay perfectly constant. Rather than depending on some particular perspective, it must remain the same value for absolutely anyone measuring it.

Newton's laws of motion, on the other hand, require that the speed at which an object appears to move depends on the speed of the person doing the observing. If a person moves faster and faster, anything traveling in the same direction ought to appear to be moving relatively slower and slower. In the limiting case of a watcher moving at exactly the same speed as what he is watching, from his own point of view he should observe it to be at rest.

Consider, for example, two moving walkways operating along side each other in the same direction at identical constant speeds. If a woman steps on one of the walkways at the same time her husband boards the other, they would appear to each other not to be moving at all. They would be able to interact with each other as if they both were standing together on a solid rock. If the wife decides to step from one walkway to the other, it would seem to her like walking on solid ground. Although Newtonian mechanics mandates that this is the case for all types of motion, Maxwell's theory of luminous movement says nothing about it.

In 1887, a clever experiment by the physicists Albert Michelson and Edward Morley brought these dilemmas to the forefront. Using a type of apparatus suggested by Maxwell, they tried to measure the effects of Earth's motion through the "aether wind" on the speed of light. Their device compared the velocities of light waves moving identical lengths in two perpendicular directions. Because these two paths were oriented differently with respect to Earth's movement through space (and the aether through which light was thought to travel), Michelson and Morley expected two distinct values. They predicted a difference in line with the Newtonian concept of relative motion. To their amazement, they found no difference between the two measured values of the speed of light, demonstrating that observers' motions had no effect. Therefore, unlike any other natural phenomena known at the time, light's motion seemed to shirk Newton's laws. This is one of the famous negative results in scientific history—an experiment that has been repeated again and again with the same outcome.

In 1892, the Dutch physicist Hendrik Lorentz and the Irish physicist George Fitzgerald independently tried to resolve this contradiction by proposing that objects contract along their direction of motion due to the pressure of the "aether wind." They hypothesized that the Michelson-Morley apparatus shrank slightly along the direction of Earth's path through space. This "Lorentz-Fitzgerald contraction" precisely negated the Newtonian effects, leaving light's apparent speed the same.

The researchers could not, however, explain why such a startling balancing act should occur. To doubters, the exact canceling out seemed too much like coincidence. Moreover, the motivation for the effect didn't make sense if light moved through a vacuum and no one indeed had detected any aether. Maxwellian physics and Newtonian physics, each with astonishing successes, seemed doomed to clash.

These unique difficulties called for someone young, with a fresh approach and little prejudice of ideas, to pull the suppositions of physics apart, rearrange them, and put them back together on sturdier bases. Not until then could a unified view of nature even be attempted.

Chasing a Light Wave

Albert Einstein was born in Ulm, Germany, in 1879 and moved with his family to Munich shortly thereafter. At sixteen years of age, while attending Luitpold Gymnasium (a high school in Munich), Einstein had already started to pursue some of the questions that would shape his life's work. One of these issues had to do with reconciling the constancy of the speed of light with Newton's laws of motion. What would happen, he wondered, if you chased a light wave, running faster and faster until you caught up with it?[4] Would it look like it was standing still? If so, how then to explain the hypothesis that light's speed should never change?

These questions occupied Einstein's mind while he studied at the ETH (the German initials for the Federal Institute of Technology, a polytechnic university in Zurich, Switzerland) from 1896 to 1900. In that intellectually stimulating climate, he found ample opportunity to discuss his ideas with bright fellow students, including Michele Besso, who became a lifelong friend and correspondent; Marcel Grossmann, who became an important collaborator as well as a generous pal; and Mileva Marić, who became his emotional soulmate and first wife.

After Einstein received his degree, he was dismayed to find that no academic positions were available for him. Luckily, Grossmann helped him obtain a job as a patent clerk in Bern, Switzerland. At the patent office, he worked efficiently, allowing him considerable time to ponder the light wave dilemma. Finally, at age twenty-six, he discovered a brilliant resolution to the problem, namely the special theory of relativity.

Einstein's theory sacrifices neither the constancy of light speed nor the notion that velocities are relative. Furthermore, it doesn't invoke a hypothetical aether. Rather, it sets aside Newton's concepts of absolute space and time, replacing them with measures that depend on observers' relative speeds.

In the Newtonian conception, space has no dynamics; it is set in stone. No matter what the circumstance, it never changes its form. Consequently, like the yard lines on a running track, it provides the fixed markers by which motion can be measured. More specifically, three-dimensional space constitutes three mutually perpendicular "running tracks": an x-axis (corresponding, say, to length), a y-axis (for width), and a z-axis (for height). One can use such a three-dimensional scale to specify exactly where any object in the universe lies. No matter where it is, its x, y, z coordinates—corresponding to its location along each of those directions—have precise, objective values.

Time, in Newtonian physics, similarly offers a fixed scale. The duration of an event should seem the same for anyone with an accurate timepiece. Seconds, minutes, and hours in one part of the cosmos should be the same for all parts. This defines a "universal clock"—designated by the t coordinate—for all regions of the universe.

Einstein found that by abolishing absolute space and time, he could reconcile Maxwell's equations with Newton's laws. He assumed instead that yardsticks and timepieces record different values depending on an observer's motion relative to what he is measuring. This elegantly eliminated all of the contradictions associated with the universal speed of light.

In special relativity, an effect called time dilation states that the faster observers travel, as they approach the speed of light, the slower their clocks move relative to those set on the ground. This precept resolves the issue of the wave-chasing runner. If a runner's clock is slowing down at the pace determined by special relativity, no matter how fast he runs, light seems to him to be moving at the same speed. Therefore, the time dilation guarantees that he can never catch up with a light wave.

To help understand this effect, picture a concert hall with ten rows of seating. A pianist sits up on stage, playing the "Minute Waltz" over and over again. For some reason, you are the only person in the audience; perhaps other guests were put off once they saw the concert program. You sit in the back row for a while, until you realize it is okay to move closer to the stage. You try out the ninth row, then the eighth, getting nearer and nearer to the pianist.

Realizing that he has an audience of one, the pianist decides to play a joke on you. Every time you move one row closer, he slows down his pace. By the time you are near the stage, he is tapping the ivories at tortoise speed.

To reduce the possibility of being caught shirking his duties—by tradition, the "Minute Waltz" must be played in sixty seconds or less—the pianist has taken precautions. He has cleverly keyed the only timepiece in the room—a grand clock above the stage—to the metronome he uses to set his own pace. Every time he slows down, the grand clock slows by an identical fraction.

You decide to gather evidence and lodge a complaint about the increasingly lethargic pace of the performance. Each time you move one row closer to the stage, you time the length of the piece. (You've forgotten your own watch and rely on the grand clock's readings for this.) Because the pianist slows down at the same pace as the clock, the piece takes identically one minute each time. Therefore, your "objective data" indicates, against your instincts, that the pianist continues to play the minute waltz at a constant rate. "Time dilation" robs you of the evidence to make your case, in the same manner that it would prevent a runner, chasing a light wave, from seeing it move slower.

Special relativity also encompasses a related effect that overturns the Newtonian idea of absolute space. Einstein's modified version of the Lorentz-Fitzgerald contraction excludes the existence of aether and proposes instead that measuring rods (yardsticks and the like) read differently for fixed and co-moving (moving along with the instrument) observers. Hence, according to this special relativistic effect, a spaceship traveling sufficiently close to the speed of light might seem 30 yards long for those on board, but only 10 yards long according to someone watching from Earth.

In 1905, which many scholars call Einstein's "miraculous year," he published a number of critical papers in physics, including "On the Electrodynamics of Moving Bodies," in which he first expounded the theory of special relativity, and "Does the Inertia of a Body Depend on Its Energy Content?" in which he demonstrated that mass and energy are equivalent (his famous $E = mc^2$). Another paper, based on his thesis, calculated the size of molecules. He also produced articles on a molecular property known as Brownian motion, and an atomic behavior known as the photoelectric effect. Any one of those papers would have won him acclaim. When he received the 1921 Nobel Prize, curiously it was for the photoelectric effect, not for relativity, although his achievements in the latter area were clearly impetus for his recognition.

The Prophet of Space-Time

In 1907, the Russian-German mathematician Hermann Minkowski, a former teacher of Einstein's, rewrote the equations of special relativity in a novel form by use of a dimension beyond the traditional three. In doing so, he unknowingly resurrected the four-dimensional concept of d'Alembert and Lagrange, and echoed the writings of Wells and the anonymous author "S." Minkowski found that by identifying time as the fourth dimension, and then merging it with space in an amalgamation called space-time, he could express the special theory of relativity more simply. Maxwell's equations of electromagnetism, he discovered, could also be written in a more straightforward manner by use of the space-time approach.

Minkowski was born in 1864 in Alexotas, Russia. At the age of eight, he moved with his family to Königsberg in Prussia (now Kaliningrad, Russia), where he received a broad classical education. It was at that point, historian Peter Galison speculated, that Minkowski likely became exposed to the writings of Plato, including the allegory of the cave.[5] The notion that our sensory perceptions are mere shadows of the truth would later feature strongly in Minkowski's ideas.

Attending the University of Königsberg, where Kant once reigned, Minkowski studied with Wilhelm Weber, who exposed him to the field of electromagnetism. This interest continued after he received his doctorate in 1885 and moved on to a teaching job at the University of Bonn. At Bonn, he became fascinated with the experimental work of Helmholtz, Hertz, and J. J. Thomson, all of whom were involved with the subject. It is possible that he read about Helmholtz's notion of two-dimensional beings on a sphere, an analogy similar to the bookworms of Gauss and Sylvester.

A clearer source for Minkowski's interest in higher dimensions lay in his appointment to a chair at the University of Göttingen, after a six-year stint at the ETH. It was at the ETH where he taught Einstein. Minkowski was recruited to Göttingen by the brilliant mathematician David Hilbert. Hilbert himself was appointed by Felix Klein, who had raised Göttingen's already considerable mathematical profile to stratospheric heights.

In 1905, Minkowski and Hilbert cotaught a series of seminars on electromagnetic theory. It was then that they became aware of Einstein's remarkable discovery. Encouraged by Hilbert, Minkowski wrestled with ways of reshaping the conceptual world to match the magnitude of special relativity's radical transformations. Immersed in the hallowed Göttingen tradition of non-Euclidean space and higher dimensional geometry, he began to think along the lines of a four-dimensional structure.

In addition to Einstein and Hilbert, Minkowski was strongly influenced by the work of Lorentz, as well as by the ideas of the French mathematician Henri Poincaré. If Einstein was the father of special relativity, Lorentz and Poincaré were the actively interested godfathers. Though basing his theories on the fictitious aether, Lorentz developed a set of transformations that made Maxwell's equations yield the same results for moving as well as fixed perspectives. Poincaré realized that the Lorentz transformations were rotations in a four-dimensional space, if one set the fourth coordinate to be an imaginary (square root of negative one) multiple of the time.

Interestingly, Lorentz and Poincaré each came tantalizingly close to discovering special relativity before Einstein. Lorentz's reliance on the aether precluded him from banishing absolute space and time. Poincaré similarly believed in the aether, but stressed the role of geometric transformations of space from one perspective to another. He also emphasized the connection between dimensionality and physical sensation, maintaining the theoretical possibility of training the eyes to perceive the fourth dimension. In his 1903 book, *Science and Hypothesis,* he imagined someone subjected to different optical stimuli: "A being educating his senses in such a world would no doubt attribute four dimensions to complete visual space."[6]

Nevertheless, Minkowski was much bolder than Poincaré in his pursuit of the physical possibility of four dimensions. Unlike Poincaré, he became convinced that the four-dimensional realm was indeed the truth, and that the sense of three dimensions was a complete illusion. Poincaré, on the other hand, wrote that "the language of three dimensions seems the better fitted [than four] to our description of the world."[7] Moreover, though Poincaré spoke of space and time as separate entities, Minkowski reached the conclusion that they were different aspects of the same thing.

A Perfect Union

Minkowski announced his findings in a famous public lecture given in Cologne in 1908. The speech began in an unusual manner for a physics talk: "The views of space and time which I wish to lay before you have sprung from the soil of experimental physics and therein lies their strength. They are radical. Henceforth space by itself and time by itself are doomed to fade away into mere shadows, and only a kind of union of the two will preserve an independent identity."[8]

Indeed, Minkowski's four-dimensional reformulation of physics had an electrifying effect on the progress of physics. It provided a new array of mathematical tools by which scientists can describe and analyze physical occurrences. These served to simplify relativity's goal of being able to view situations from a wide range of perspectives, including frameworks moving at various speeds with respect to each other.

The playing field of Minkowski's conception of relativity is called a *space-time manifold*, also known as the *continuum*, the set of all things at all times. Essentially, the space-time manifold encompasses the universe itself, as far back into its past and as far forward into its future as one can imagine. It includes anything that has ever happened or will ever happen through eternity.

Instead of spatial points, the basic units in Minkowskian relativity are space-time *events*. Each event represents the location and time of a physical occurrence, as characterized by four numbers, known as its *space-time coordinates*. These consist of three numbers representing spatial position—x, y, and z—as well as one number representing the time—t. One can chart these on a *space-time diagram*, a four-dimensional map plotting relevant events according to their coordinates.

For instance, if one wishes to specify the birth of Einstein using space-time coordinates, one can supply the longitude, latitude, and height above sea level of his birthplace (Bahnhofstrasse 135 in Ulm), as well as the time of his delivery (11:30 A.M. on March 14, 1879). One can then plot this as a point on a space-time diagram, with perpendicular axes delineating x, y, z, and t. Because of the physical impossibility of depicting such a four-dimensional graph, one typically chooses two or three of these axes for the actual plot—x versus t for instance.

Physics often considers pairs of occurrences—starting and stopping points—rather than single instances. In that case, one can plot each event on a space-time diagram and examine the change in spatial and temporal displacement between beginning and end. This can be represented on the graph by a line segment linking the points, capped by an arrow showing the direction. Such a mathematical entity is known as a *four-vector*.

For example, suppose Hans throws a football to his friend Peter. Peter is standing 20 feet to the east of Hans and 30 feet to the north, on top of a hill that has a height of 10 feet. Peter catches the ball two seconds later. Then the four-vector representing the displacement of the football from Hans's perspective has coordinates 20 feet, 30 feet, 10 feet, and 2 seconds. These four numbers are also known as the vector *components*. One would plot the four-vector as a line segment linking the space-time point representing Hans with that depicting

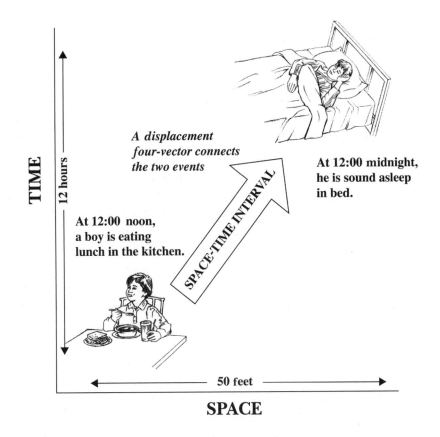

A displacement four-vector connects the two events

At 12:00 midnight, he is sound asleep in bed.

At 12:00 noon, a boy is eating lunch in the kitchen.

SPACE-TIME INTERVAL

TIME

12 hours

50 feet

SPACE

Shown here is a simple example of a space-time diagram. Depicted are two separate events: a boy eating in his kitchen at noon, and the same boy sleeping in his bedroom at midnight. These events are linked by a four-vector, the length of which is known as the space-time interval. It represents the shortest distance through space-time between the two points. The space-time interval can be positive, negative, or zero, depending on the positions and times of the events under consideration. To view this diagram from another relativistic perspective (from a spaceship, say), one need only rotate the axes.

Peter. An arrow placed on Peter's end of the line segment pointing toward him would indicate the direction in space-time that the football travels.

There are other four-vectors besides those representing displacement. For example velocity and acceleration are four-vectors as well, each possessing four components. In general, four-vectors have magnitudes (lengths) and directions (the way the arrows are pointing).

Not all physical entities can be expressed as four-vectors, however. Some quantities such as temperature don't point in any direction. It would be silly for a weather forecaster to say it's going to be a northwesterly 30 degrees outside. The wind velocity could shift in such a manner, but not the temperature itself. Therefore temperature is an example of a directionless mathematical object known as a *scalar*. Scalars don't have components, they are just single numbers.

A third type of entity, the workhorse of the field of differential (calculus-based) geometry, is called a *tensor*. A tensor is essentially a rule sheet, or rulebook, delineating a particular way of transforming one mathematical object into another. For instance, a tensor applied to a four-vector might render it a different four-vector, or, alternatively, a scalar. Not all rules are possible; mathematicians restrict these to certain kinds of transformations. Nevertheless, tensors offer a great deal of flexibility, rendering them essential tools for modern geometric description.

Tensors are often depicted in arrays, with rows and columns like a spreadsheet or a checkerboard. For example a tensor transforming one four-vector into another would be a four-by-four checkerboard. Each column represents one of the dimensions of the old four-vector (*x, y, z,* or *t*), and each row, one of the dimensions of the new four-vector. Placed in each square of the checkerboard is a function (mathematical rule) showing the part the old vector components play during their transformation into the new. By applying these rules, the tensor acts as a well-programmed machine, precisely carrying out its transformational duties.

One of the most important operations in Minkowskian relativity involves a four-vector, a scalar, and a tensor. Suppose one wishes to measure the "distance" between two events in space-time, for example, the amount of space-time traversed by Hans and Peter's football. This differs from the conventional spatial distance, calculated by use of the Pythagorean theorem. According to the Pythagorean theorem, the square of the distance between two objects is the sum of the squares of their x, y, and z coordinate differences. But Minkowski space-time is not exactly a Euclidean geometry, so one must amend this formula.

Replacing ordinary distance is the concept of the *space-time interval*, the length of the shortest path between two events. Mathematically, this is the magnitude of the four-vector representing space-time displacement. To find this, one must modify the Pythagorean theorem by adding the squares of the spatial coordinate differences, then subtracting the square of the time difference. The result is the square of the space-time interval.

To accomplish this feat, one can use a special rule sheet called the *metric tensor.* Recall that a metric expresses the rules for computing distances in a particular geometry. What better place to house these formulas than in a tensor, geometry's ultra-flexible transformational machine? By placing the proper expressions—namely the modifications to the Pythagorean theorem that change plus to minus for time—in the appropriate squares of the metric tensor's four-by-four checkerboard, one can convert any four-vector into its space-time interval.

The space-time interval has very special properties. For one thing, unlike the distance, it can be zero or even negative for two separate events. An interval of zero means that the two events lie on the same path through space-time that a light ray would take between them. A negative interval, or timelike separation, means that the events can communicate more slowly than light would take (the football toss, for instance). A positive interval, or spacelike separation, indicates that the events couldn't normally interact with each other (two phones simultaneously ringing on opposite ends of the galaxy could be chalked up only to coincidence, not cause and effect, because the signal couldn't instantly jump from one to the other).

Another special property of the space-time interval is that Lorentz transformations do not change it. In this context, a Lorentz transformation embodies the Einsteinian rules for time dilation and length contraction. Dilate the time and contract the space all you wish between two events, but as long as you follow Einstein's special relativistic formulas, the space-time interval between them stays the same.

A good analogy to this is a wheel of chance at a carnival. Imagine drawing a radial line from the center of the wheel to one of the numbers (the 13, let's say). No matter how much one spins the wheel, and where the lucky number ends up, the length of the radius stays the same. One could bet on its invariance, and come up a sure winner.

This suggests an elegant way of depicting special relativity. Represent the starting and stopping points of any observed process as events in a space-time diagram. Draw the four-vector between the events, situating the tail at the origin (zero point) of the diagram. Then rotate the diagram about the origin in the plane spanned by the time axis and the space axis under question (the one in which the relativistic motion takes place). This is like spinning a wheel of chance. What happens is that the space and time components of the four-vector shrink or grow accordingly, depending on where the spinning wheel ends up, but the space-time interval (the wheel's radius) stays the same. As the spatial part decreases, leading to Lorentz-Fitzgerald contraction, the temporal portion increases, producing time dilation—the effects balancing each other to maintain the interval's status quo.

The varied positions of the displacement four-vector as it rotates through space-time constitute all possible observational viewpoints, moving at fixed velocities relative to each other. For example, in the case of players Hans and Peter, one position of the four-vector represents the boys' perspective. Another depicts the vantage point of a flea that happens to sit on the football. Yet another denotes the window view of a businessman whizzing by on the Göttingen Express. Like a slide cartridge turning in a rotary projector, every angle frames a slightly different relativistic picture, casting varying images of space and time.

Not only did Minkowski encompass special relativity in one basic scheme, he also found a way of reducing Maxwell's four equations to one. By defining quantities known as the electromagnetic field tensor and the four-current, he discovered a simple relationship that describes the vast range of electric, magnetic, and luminous phenomena. The electromagnetic field tensor contains all of the components of the electric and magnetic fields, written in the form of a four-by-array. The four-current is a four-vector that, instead of displacement, describes the electric current as well as the electric charge. Minkowski's succinct formulation of Maxwell's equations states that the four-current is a function (technically known as the divergence times a constant) of the electromagnetic field tensor. The entire spectrum of light and the full scope of electricity and magnetism are all expressed in an equation that can be written the size of one's thumb; what a giant step for unity.

Superfluous Learnedness

Despite its elegance, for several years after its proposal, Einstein was not impressed by Minkowski's treatment of his theory. Whenever possible, Einstein preferred tangible treatments of physical problems, and saw Minkowski's approach as overly abstract. When Einstein was a student at the ETH, he missed many of Minkowski's classes and was reprimanded by the college's director for nondiligence.[9] Minkowski said he was "brilliant . . . but a lazy dog."[10] He preferred spending time in the physics labs, seeing how nature really worked, rather than paying attention to mathematical abstractions.[11] Like a carpenter listening to an artist rave about the aesthetics of wood, he was unable to see how higher math had any physical application. No wonder that when Minkowski, with his geometrical hocus pocus, transformed special relativity into a four-dimensional concoction, Einstein found little point to the exercise. In fact, Einstein was concerned that the use of terms such as the "fourth dimension" would generate only confusion, a "superfluous learnedness"[12] as he put it. Now that the mathematicians had taken over relativity, he bemoaned, he could barely understand it himself.[13]

As the historian John Stachel has pointed out, Einstein's initial apprehension about the fourth dimension was such that in 1908, he and his colleague Jakob Laub essentially rederived Minkowski's four-dimensional results on electrodynamics in three-dimensional form, stating that Minkowski's "work makes rather great demands mathematically on the reader."[14]

Several months after Minkowski delivered his influential talk, his appendix ruptured suddenly. He was only forty-four when he died. Reportedly, on his deathbed he regretted having to pass away when relativity was still in its infancy.[15]

Somehow in death, Minkowski wielded greater influence on the ideas of his former student. By 1910, Einstein was starting to get more accustomed to the four-dimensional space-time approach. He wrote in a letter to the physicist Arnold Sommerfeld, who had extended Minkowski's work into an algebra of four-vectors: "The consideration of the formal relations in four dimensions seems to me an advance. . . . I have probably expressed myself wrongly in this respect when I [previously] talked to you."[16]

In his lectures, Einstein started to refer to Minkowski's work as a more elegant expression of special relativity. In a talk on relativity in January 1911, Einstein discussed "the highly interesting mathematical elaboration that the [special relativity] theory has undergone, thanks mainly to the sadly so prematurely deceased mathematician Minkowski. . . . The further pursuance of this formal equivalence of the space and time coordinates in the theory of relativity . . . makes its application substantially easier."[17]

Two main factors changed Einstein's mind about the use of the fourth dimension. First, he witnessed its increasing incorporation into descriptions of special relativity—Sommerfeld's work, for example, as well as a 1911 textbook on the subject written by Max von Laue. Einstein considered von Laue's book—the first ever written about relativity, offering a decidedly four-dimensional approach—"a little masterpiece."[18]

Second, Einstein began to ponder the role of four-dimensional geometries in helping to expand special relativity into a comprehensive theory of dynamics. Special relativity only considers motion at constant speeds; it does not address the case of acceleration. Describing accelerated motion, however, is such a key element of dynamics that a theory without it is like a course in auto mechanics that fails to mention how engines work. Moreover, acceleration plays a critical role in discussions of gravitation, which Einstein eagerly wanted to characterize in relativistic terms. To construct a complete description of motion, including accelerated frameworks, Einstein came to realize that he needed to employ the power of higher mathematics, including the four-dimensional approach he had previously disregarded.

Portraits of the Fourth Dimension

Like a convert to a new religion, Einstein struggled adamantly against misinterpretation of the doctrine. To head off mystical interpretations of the concept, he began to insist that there was nothing extraordinary about the fourth dimension—a theme that would appear in all of his popular writings on the subject. For example, much later, he and Leopold Infeld would write: "Our physical space as conceived through objects and their motion has three dimensions, and positions are characterized by three numbers. The instant of the event is the fourth number. . . . Therefore, the world of events forms a four-dimensional continuum. There is nothing mysterious about this, and the last sentence is equally true for classical physics and the relativity theory."[19]

Yet all around Einstein, new movements in art seemed to suggest that the fourth dimension was something revolutionary and bizarre. Cubism, a style pioneered by Pablo Picasso and Georges Braque while working in a Paris artist colony starting in 1907, aspires to depict scenes from all manner of spatial viewpoints. Shunning the Renaissance notion, maintained for centuries, of presenting objects as they would look from single vantage points, it attempts to show simultaneously their fronts, backs, sides, tops, and bottoms. In sharp contrast to the techniques introduced by Giotto and perfected by Leonardo, the perspective lies outside the world itself. Cubist paintings have an extra-dimensional quality similar to A Square's view when the Sphere took him on a scenic tour of Flatland's landscape. Max Weber's *Interior of the Fourth Dimension* (1913), a wondrous vision of the modern city, exemplifies the close ties between Cubism and the concept of hyperspace.

What Cubism did for space, Futurism would do for time. The Futurists were a radical group of artists based in Italy who believed that still images were outmoded in the age of cinema. Rather than trying to capture solitary snapshots of reality, they aimed at presenting expositions of continuous motion. Like consecutive frames of a motion picture displayed all at once, their work renders spatially what the eye normally observes over time. For example, Giacomo Balla's 1912 painting *Dynamism of a Dog on a Leash* depicts a dog with a jumble of mouths, ears, legs, and tails, blending the various positions all these parts would have during a typical walk.

A very independent-minded French artist, Marcel Duchamp, bridged aspects of Cubism and Futurism in his early work. His masterful *Nude Descending a Staircase* (1912) portrays a feminine form captured over time from various angles. At first glance, the painting seems a blur of scratchy lines and overlapping shapes. Then, somehow, the piece seems to take on a life of its own, like the mysteriously moving portraits in horror movies. In an odd sense, it is anatomy captured on a space-time diagram, as if Minkowski moonlighted as a medical documentarian.

Duchamp's most intricate work, begun in 1915 and completed in 1923, was a conscious ode to higher dimensions entitled *The Bride Stripped Bare by Her Bachelors, Even* (*The Large Glass*). (No, not all of Duchamp's pieces were about nudity.) It is a whirl of motion, showing the gyrating machinery of seduction. Uniquely, its images are presented on a clear pane of glass, rather than on canvas. Duchamp hoped that it would present a looking-glass vision of reality, like Alice's view before she stepped through the mirror. The glass shattered before the piece was installed. Reportedly Duchamp was delighted, not mortified, commenting that the weblike cracks improved upon the piece.

Many historians used to argue that Cubism, Futurism, and related styles were direct responses to Einsteinian relativity. However this attitude has largely changed with the publication of Linda Dalrymple Henderson's meticulously researched book *The Fourth Dimension and Non-Euclidean Geometry in Modern Art*. Henderson demonstrates that the artists of that period were unfamiliar with Einstein's findings and the notion of space-time. As she points out, their concept of the fourth dimension likely stemmed from nineteenth- and very early twentieth-century sources that predated Einstein's work, including the imaginative writings of Hinton, Abbott, and Wells; graphical depictions by Stringham and others; popular articles and speeches by scientists and mathematicians such as Newcomb, Sylvester, and Helmholtz; the spiritual beliefs of Theosophy and other occult movements, and the like.

The lack of public awareness of the writings of Einstein and Minkowski during the years before World War I is reflected in a collection of popular essays about the fourth dimension written in 1909. The essays were submitted to *Scientific American* in competition for a $500 prize offered by an anonymous donor. Many of the responses were quite creative. Covering a wide range of philosophical territory, various pieces referred to ideas by Kant, Gauss, Hinton,

Zöllner, and others. Yet not a single essay mentioned relativity, space-time, Einstein, or Minkowski. Not for another decade would their research be widely known beyond the physics community.

In the case of Duchamp, cultural historians such as Linda Henderson and Gerald Holton have traced his awareness of the fourth dimension back to the popular writings of Poincaré and another French mathematician named E. Jouffret. Jouffret wrote a 1903 treatise on four-dimensional geometry that Duchamp had read with keen interest. The way historians discovered this was literally through a paper trail. As Holton relates, "He had evidently studied Jouffret. It says so in one of his notes. His habit was to write notes and then put them in a box. These were scraps of paper which later on were then put together and published as books."[20]

The unfamiliarity of modern artists with Einstein's work was reciprocated with sheer disinterest. As innovative and cultured as Einstein was, he could not see the merit in the newfangled styles. Throughout his life, his strong preference was traditional art and classical music. Moreover he balked at comparisons between modern paintings and his own scientific creations. Referring to Cubism he once said, "This new artistic 'language' has nothing in common with the Theory of Relativity."[21]

Indeed, during the heyday of Cubism and Futurism, Einstein was busy shaping his own masterpiece. Gradually he was putting together his relativistic theory of gravitation, the pinnacle of his scientific achievement. And like Leonardo and Michelangelo at their most creative, little did he rest until it was completed.

CHAPTER 4

Getting Gravity in Shape

Every boy in the streets of our mathematical Göttingen under-stands more about four-dimensional geometry than Einstein. Yet, despite that, Einstein did the work and not the mathematicians.

—DAVID HILBERT, leader of the Göttingen school of mathematics

The Happiest Thought

Einstein's construction of the general theory of relativity, his Sistine Chapel of gravitational principles, began in fits and starts. Around the same time Minkowski proposed the notion of space-time, Einstein had "the happiest thought of [his] life."[1] Strangely, this happy thought involved picturing someone falling off the roof of his house. Einstein imagined a man losing his footing and then dropping to the ground. While falling, he lets go of something he is holding, and notices (assuming no air resistance) that it remains next to him, as if at rest. Actually, it freely falls at the same rate—an effect Galileo reportedly first discovered by dropping objects of different weights.

The originality of Einstein's thought is embodied in a precept called the *equivalence principle*. No experiment, it states, should be able to distinguish between an object freely falling due to gravity and the same object being at rest. For example, two girls playing a game of marbles while sitting on a free-fall ride in an amusement park should not be able to tell the difference from playing the game on the sidewalk outside their house.

Einstein realized that the equivalence principle was the key to developing a relativistic theory of gravity. In special relativity, two runners proceeding at identical speeds can behave as if they are at rest with respect to each other, keeping up a steady conversation if they wish. In a gravitational field, the same is true, but only if the runners are both dropping toward Earth at the same rate; for instance, if they shared a freely falling elevator (large enough to have a running track inside, that is). Therefore, a free-falling frame precisely mimics the conditions of special relativity, for which Minkowskian space-time offers an exact description.

This realization set Einstein on a course toward designing a representation of gravity. By sewing together myriad free-falling frames, like patches on a quilt, he surmised that he could fashion the complete gravitational dynamics of the universe. But what thread, he wondered, could create a seamless result—with each frame connected perfectly to the next?

From 1908 to 1911, while holding successive university positions in Zurich and Prague, Einstein experimented with a number of rudimentary versions of a relativistic theory of gravitation. None of them worked. It was then that Einstein realized that he needed to carry out his design by use of a more sophisticated mathematical framework, including manipulations of four-dimensional space-time geometries. At that point, however, he lacked the mathematical expertise to complete his task.

Einstein's theoretical efforts were interrupted in early 1912 by more practical considerations. Lorentz, stepping down as director of the physics department of the University of Leiden in Holland, offered Einstein a prestigious position as his replacement. Einstein respected Lorentz enormously, viewing him as a scientific father figure. He felt indebted to Lorentz for some of the physical notions that comprised the bedrock of special relativity.

Nevertheless, Einstein felt compelled to turn down Lorentz's offer. Einstein had already promised Marcel Grossmann that he would move back to Zurich and take on a position at the ETH. The Leiden position went instead to Lorentz's second choice, the

Austrian physicist Paul Ehrenfest, who would later play a prominent role in the development of theories of dimensionality.

Since their student days together, Einstein and Grossmann had maintained a close friendship. An adept mathematician, Grossmann had returned to the ETH in 1907, joining the faculty first as professor of geometry and then as dean of mathematics and physics. He was appointed dean at the tender age of thirty-three, highly unusual for the times. One of the first decisions the young dean made was to invite Einstein to join him as professor of physics. Einstein was eager to move back to Zurich because he and his wife, Mileva, had strong ties to the city.

When Einstein joined the faculty of the ETH, he resumed work on relativistic gravitation but still couldn't get the mathematics right. Desperate to complete his theory, he called upon his friend for assistance. "Grossmann, you must help me, otherwise I'll go crazy!" Einstein reportedly implored.[2] Grossmann, skeptical of physics but always eager to lend a hand, introduced Einstein to the bizarre world of Riemannian higher-dimensional geometry, with its warped, non-Euclidean vision of space. Einstein found Grossmann's assistance just what he needed to develop further the mathematical structure of his theory.

Einstein's stay at the ETH turned out to be short lived. In 1913, he was offered a special triple appointment in Berlin as researcher at the Prussian Academy of Sciences, professor at the University of Berlin, and director of the yet to be established Kaiser Wilhelm Institute for Physics. Although he enjoyed living in Zurich, the Berlin offer, involving no teaching commitment whatsoever, was too generous for him to resist. Feeling close to completing his gravitational theory, Einstein relished the idea of having unlimited time to do his research. Enthusiastically accepting the invitation, he, Mileva, and their two boys moved to Berlin in 1914.

There was another, more personal, reason for Einstein's decision to leave Zurich. Around that time he had grown increasingly dissatisfied with his marriage to Mileva, whom he found "very cold and mistrustful."[3] His heart turned more and more to his warmhearted first cousin, Elsa Löwenthal, who lived in Berlin and made it clear that she cherished him. After arriving in Berlin, Einstein became even closer to her. She even began to give him "motherly" advice about his grooming, which his intense concentration had caused him to neglect. Needless to say, Mileva was not very happy. Assisted by Michele Besso, she took the boys and returned to Zurich, indignant by what she saw as her husband's neglect. Einstein moved into an apartment by himself, working ever harder on his theory of gravitation.

One of Einstein's incentives for finishing his theory was the realization that other physicists were working on the same problem, notably Gustav Mie in Greifswald and Gunnar Nordström in Helsinki. However, as Einstein soon discovered, the models of Mie and Nordström lacked elements required for a successful relativistic theory of gravitation. Mie's approach did not obey the equivalence principle. Nordström's comprised a special relativistic theory of gravity within the context of Minkowski space-time—resembling a scalar approach Einstein had discarded in 1911. In a 1914 paper on the subject, Einstein, along with the physicist Adriaan Fokker, pointed out the shortcomings of Nordström's model and suggested improvements.[4]

In 1916, Einstein finally announced success, publishing the general theory of relativity in its complete form. As he wrote to Besso, he felt "satisfied but rather worn out."[5] His intense work on the project had taken its toll on his health, which would deteriorate in the years to come in the form of liver and stomach ailments. It was the unfortunate occupational hazard of someone whose life's mission to unravel universal mysteries superceded his own personal comfort.

For the sake of science, at least, the effort proved well worth it. The general theory of relativity is an extraordinarily elegant set of equations that describe gravity by relating the mass and energy in any region of the universe to the structure of space-time in that sector. Unlike Newtonian gravity, which ties together spatially distinct masses with an invisible "thread," Einsteinian gravitation operates purely locally. Thus it resembles another highly successful theory, Maxwell's equations of electromagnetism, in that all the action relating to a single point in the universe takes place at that point itself.

The mechanisms of the general theory are like an assembly line in which raw material is transformed, over a multistep process, into a tapestry of fine fabric. Here the material is the matter and energy of a region of the universe, and the fabric is that of space and time itself. The process starts with an equation that relates two tensors (rule sheets for how to process mathematical objects), called the stress-energy tensor and the Einstein curvature tensor. The stress-energy tensor describes the material properties at a given point; for example, does that region contain compact stellar matter, a trace of a lightweight substance, a burst of radiation, or is it just empty? The Einstein curvature tensor, itself the sum of two other tensors, conveys geometric information about the shape of space-time at the same location. Unlike Minkowski's model, which has three fixed perpendicular directions and time—arranged like a four-dimensional box—the general relativistic universe features unlimited ways it can be stretched or curved. It can be shaped like a hypersphere (a four-dimensional sphere), a hyperboloid (a four-dimensional saddle), or myriad other shapes. The curvature tensor classifies each of these possibilities.

The curvature tensor, in turn, can be expressed in terms of mathematical objects known as *connections.* These describe how four-vectors (directed line segments pointing from one space-time event to another) change when they are transported parallel to themselves along various paths. In Minkowski space-time, considered "flat" (not curved), four-vectors do not alter when moved parallel to themselves along any route. In a curved space-time, on the other hand, they can change quite dramatically depending on the trajectory chosen.

To understand this effect, let's imagine twins, Marius and Darius, who each have the unusual habit of always facing northward. Even when walking east or west, they still face the northerly direction, forcing them to shuffle sideways. They can keep this up for hours, days, or months if they have to, for each is an expert walker. One day, they challenge each other to a monumental race: from Quito, Ecuador, right near the Equator, all the way to the North Pole. Each elects to take a different route.

Marius decides to walk due north. Whenever he needs to cross a body of water, his friends arrange a boat for him so that he can continue to proceed in exactly the same direction. He crosses the United States and Canada, constantly looking to the north. Finally, when he reaches the North Pole, he stops, finding himself facing in the direction of Russia.

Darius, on the other hand, decides to shuffle sideways along the equator, still facing north all the time, until he reaches the Indian

Ocean just off the coast of Sri Lanka. (Friends have provided him with boats as well for the long aquatic stretches of the trip.) By then he is halfway around the world from where he started. From there he proceeds due north, across Asia. He arrives at the North Pole, several months after his brother. But instead of facing Russia, Darius is facing Canada, exactly in the opposite direction.

Neither brother ever changes the way he is facing. Both start out in the same place and finish in the same place. Yet they end up in diametrically opposite positions. Where lies the trick? The reason is that the earth's curvature acts to change the direction of any perpendicular set of coordinates moved along its surface. For instance, north isn't always in the same direction, as visitors to either pole soon find out. (An old riddle imagines a house on the North Pole, with each side having southern exposure.) For similar reasons, a four-vector transported parallel to itself along different paths through curved space-time will have, in general, two different directions. The difference between the two of them can be quantified by use of the connections.

In the final step of the process, the connections can be expressed in terms of the components of a metric. These determine the values of all possible space-time intervals—namely the shortest distances between all pairs of points in space-time. One of the simplest cases is that of Minkowski space-time, where the metric machine processes four-vectors by means of a modified version of the Pythagorean theorem (the sum of the squares of the spatial components minus the square of the time component). More complicated scenarios, such as hyperspheres, lead to metrics with more elaborate properties, represented by more complex sets of instructions for finding the intervals associated with four-vectors. These generate more complicated structures of shortest distances between events, producing effects such as parallel lines coming together as the intervals between them shrink.

The end result of these mechanisms is that the matter and energy content of a region of the universe generates a particular web of space-time intervals between the points of that zone. This space-time web, in turn, guides the motion of objects in that domain. Alternatively, as John Wheeler has succinctly put it, "Space-time tells matter how to move and matter tells space-time how to curve."[6]

Thus Einstein's epic voyage that began with his "happiest thought" about falling bodies culminated in a radically new concept of gravity. Einstein replaced the Newtonian notion of masses exerting distant tugs on each other with a more local concept: the idea that space-time is flexible and that its curvature serves as a conduit of gravitational force.

The Lost Proposal

It was not enough for Einstein to fashion a mathematically adept theory. He had to defend it against competitors by providing means of experimental verification. He suggested several key tests, including an explanation of the orbital procession of Mercury (how Mercury's path around the Sun advances over time) and a prediction concerning the bending of starlight by the Sun's gravitation.

One of the theories Einstein's competed with was Nordström's. As a scalar, not a tensor, approach, Nordström's model offered vastly different experimental predictions. For example, light rays wouldn't be deflected by gravity. Both Einstein and Nordström were curious to see which approach would turn out to be correct. Nordström graciously conceded even before the answer became fully known. It was not hard for him to move to the other camp. After all, since his student days, Einstein was one of his idols.

Born in 1881, Nordström began his career as an engineer, then aspired to be a chemist. He changed his mind again while studying chemistry at the University of Göttingen. During his one-year stay, starting in April 1906, he encountered Minkowski's extraordinary ideas and decided to switch his field to theoretical physics. Returning to Finland, he began publishing papers on the subject of special relativity, while lecturing at the University of Helsinki. He also wrote the first popular article in Finnish about Einstein and Minkowski's work.

Nordström's university colleagues were baffled by his new interest. No one else was familiar with relativity. As a dolphin in a desert, the only chance he had for free discourse was to head for the scientific oceans of central Europe. But travel funds were hard to come by.

To one such request, he received this reply: "One can study the fourth dimension at home, without any trips abroad."[7]

Interest in Einstein's attempts to encompass gravitation within relativity inspired Nordström to propose his scalar model, which he honed for several years. In 1913, he managed to find the money to visit Einstein, then at Zurich. Although Einstein didn't believe in Nordström's approach, he respected it enough at that time to present it at a scientific meeting as a solid alternative to his own work.

The following year, Nordström literally added another dimension to his theory. Five years before Kaluza, he invented the idea of higher-dimensional unification of gravitation and electromagnetism. Built on flawed foundations—an erroneous view of gravity soon overpowered by general relativity—his unified model could not support itself. Nordström abandoned this line of reasoning soon thereafter, effectively guaranteeing it little notice. Einstein, for example, never cited it, possibly because he had decided at that point to disregard theories based on Nordström's gravitational approach, or maybe because of "less than cordial" relations between him and Nordström.[8] While Kaluza and Klein would later receive ample references for their five-dimensional theories, Nordström's would virtually be ignored, until in 1982 Pais mentioned it in his biography of Einstein, and then in 1987 it was translated into English by the physicist Peter Freund. Freund's translation appeared in a respected compilation of articles related to Kaluza-Klein, elevating general awareness of Nordström's contributions.

Nordström's paper, entitled "On the Possibility of a Unification of the Electromagnetic and Gravitation Fields," was as boldly innovative as the unified approaches of Riemann and Clifford. In it, he combined Maxwell's equations of electromagnetism, written in Minkowski's form, with his own scalar theory of gravitation in a flat space-time of five different axes. Then he casually laid out the claim: "The interpretation of [these] equations . . . shows that we are entitled to regard the four-dimensional spacetime as a surface in a five-dimensional world."[9]

Nordström's five-dimensional world is a strange one indeed, because those of us living in its four-dimensional space-time would be mathematically precluded from noticing anything beyond our "surface." This condition would trap us on a kind of Flatland. We would, however, be able to make predictions about the effects of gravity on light. For those, Nordström's world would scarcely resemble the Einsteinian universe in which we actually exist.

Shortly after the final version of general relativity appeared, Nordström was presented with an exciting new opportunity—a research fellowship at the University of Leiden. Going to Leiden offered him the chance to be part of the physics mainstream and consult with the many other outstanding researchers drawn there by Paul Ehrenfest.

Good Things Come in Threes

When Ehrenfest replaced Lorentz as director of the University of Leiden physics department, he saw it as his holy mission to promote intensive, diverse discussion about the fundamental issues of physical theory. He constantly invited—even begged—guest speakers to come to a colloquium series he developed, where he would barrage them with questions. He was "the world's champion questioner in physics,"[10] pointing out with absolute candidness any virtues or flaws he saw in their lines of reasoning. He would ask them to sign a speaker's wall—now carefully preserved, a new one still being used today. To stimulate discussion, he also tried to assemble the most accomplished group of visiting scholars he could find—including Einstein, on many occasions.

Ehrenfest lived in a large yellow stucco house, just across the river from the university and the center of town. Stylistically it was completely different from the narrow brick houses around it, befitting the iconoclastic thinker inside. There he had ample room to pamper his guests and involve them in intellectual discussions.

Nordström arrived in the summer of 1916, brimming with enthusiasm. He and Ehrenfest found joy in discussing electromagnetism, relativity, and other topics of common interest. On warm evenings they would sometimes sit outside in the garden, a pleasant retreat with a bubbling stream, and think up new approaches to problems.

Only two weeks into Nordström's visit, Ehrenfest caught his "dimension fever" and began to look at an intriguing question.

*Paul Ehrenfest's spacious house in Leiden, where he hosted Einstein,
Klein, Nordström, and other noted physicists, has an architectural
style distinct from the other houses in the neighborhood.*

Which aspects of nature are specific to three spatial dimensions (plus time), and which are more general? His investigation into the matter began with this journal entry: "In Newton's three-dimensional space, planetary orbits are closed. What about non-Euclidean spaces?"[11]

Over time, Ehrenfest and Nordström discussed the features of higher-dimensional versions of Maxwell's equation, gravitational theory, and other aspects of physics. Like Abbott and Hinton, they enjoyed envisioning what the world would look like if it had many more directions of movement. With each question came a new realization about what makes our three-dimensional abode distinct from other possibilities.

Ehrenfest's fascination with the properties of space derived from his training as well as his marriage. He studied for a time at the University of Göttingen, where he was taught by brilliant minds such as David Hilbert and Felix Klein. He also enjoyed exchanging ideas with his wife, Tatyana, a mathematician interested in geometry who wrote about the teaching of geometry in schools and universities.[12]

They had met at a Felix Klein lecture and had continued their inter-play ever since. Together they worked hard to impart their mathe-matical interests to their children as well.

The Ehrenfest kids were an unusual group. Home-schooled and exposed to intellectual thinking at the youngest age, they used to have fun by playing the game "colloquium" where they gave their own "lectures."[13] The eldest daughter used to keep her dolls in a cardboard model of a hyperboloid.

These feats impressed one special houseguest at least. Einstein, a dear friend of the Ehrenfests, relished the times he could come and stay with them. He was always made to feel extremely welcome. Whenever he visited, he enjoyed playing the role of the children's "uncle," basking in the warmth of a family life that eluded him. John Stachel has remarked, "If Einstein had a father figure in his life it was Lorentz. If he had a brother figure it was Ehrenfest."[14]

It gave Einstein a special joy to report to his "father" and "brother" his completion of a four-dimensional relativistic theory of gravity. In September 1916 Einstein journeyed to Leiden, a difficult trip to get permission for at the height of World War I. Staying with the Ehren-fests, they drove out together to visit Lorentz. "Lorentz was smiling at Einstein," Ehrenfest later reported, "much as a father would regard his beloved son." Einstein returned to Berlin with a supreme sense of satisfaction.

By May 1917, Ehrenfest started to resolve some of his own ques-tions about the role of dimensionality in science. The result was a paper entitled "In What Way Does It Become Manifest in the Fun-damental Laws of Physics that Space Has Three Dimensions?"[15] It is a remarkable article in many ways, written by an astute physicist as eager as Einstein to probe nature's deepest questions. As Ehrenfest biographer Martin J. Klein has remarked, "The style is utterly dif-ferent from most physics papers. When you read it you feel it is from a live person."[16]

Ehrenfest begins the article by stating the infinite range of possi-bilities for the dimensionality of space and asking on what physical basis can we distinguish one from the other. What's the physical difference, for example, between a three-dimensional space and a seven-dimensional space?

He then lists a number of examples involving gravitational and electromagnetic forces, and considers what their behavior would be like in other than three spatial dimensions. The first issue he examines is the stability of planetary orbits. Working out the motion of planets under the influence of a star's gravity, he demonstrates that only in three dimensions are closed, stable orbits possible. In higher dimensions, planets would either spiral into their suns in fiery collisions, or keep moving farther and farther away as frozen bulks. In either case, lacking a steadily orbiting planetary home, we wouldn't be here. This argument is similar to Kant's gravitational explanation for the dimensionality of the world.

Another issue Ehrenfest considers is how dimensionality affects atomic stability. It concerns Danish physicist Niels Bohr's model of the atom, proposed in 1913. According to Bohr's theory, electrons orbit the nucleus at various distances, much like planets circle the Sun. Unlike the solar system, however, electronic orbits have discrete spacing between them; nothing can lie in between. If an electron jumps down to a lower orbit, it must do so by a special fixed amount, called a quantum of energy. The lowest energy level an electron can reach is called the *ground state.* For a given atom, it normally lies a fixed distance from the nucleus, precluding electrons from ever dropping to their demise and insuring atomic stability.

Ehrenfest found that the Bohr model behaves quite differently for other than three spatial dimensions. In that case, he calculated that there is no minimum energy level that provides a haven in which electrons can maintain steady orbits. Without such a ledge, all the electrons would drop down into the nucleus, and no matter would be stable. Rapidly, for other than three dimensions, everything in the world would disintegrate.

Moreover, even if material objects could somehow exist in higher-dimensional space, they wouldn't be able to communicate using light signals. Electromagnetic waves, and all other waves for that matter, would not be able to travel steadily in pulses. Instead, they would disperse, without conveying any information. Therefore, in short, existence would be brutal, solitary, and incredibly brief.

Ehrenfest's insightful article did not generate much response in the physics community at the time, as measured by scholarly citations and other references to the work. Perhaps its relative obscurity can be attributed to the fact that wartime conditions, as if plunging the earth into a higher-dimensional chaos, reduced communication among scientists. Or maybe the subject seemed too philosophical. Nevertheless, by writing this paper, Ehrenfest established himself as an expert on dimensionality, a subject in which he would later advise Einstein and other scientists.

A Strained Symphony

Einstein's visit to Leiden provided welcome respite from a rather isolated existence in Berlin. In contrast to the ETH, there was less opportunity for collaboration and little need for him to report daily to his offices, at either the university or the academy. Consequently, he often worked at home, donning old sweaters, smoking pipes and cigars, taking breaks to indulge his hobby of playing the violin, and making himself as comfortable as possible. But then, in 1917, comfort was nearly impossible when he grew gravely ill with digestive upsets, liver problems, gallstones, and an ulcer. During the winter of 1917 and spring of 1918, he was completely bedridden. While recuperating from the ailments, Elsa took good care of him, cooking special meals to aid his digestion. Thankful for all the affection, Einstein began to plan out divorce from Mileva and marriage to Elsa. These hopes would be realized the following year.

While devising plans for completing his personal life through union with Elsa, he also began to muse about ways of completing gravitational theory through union with electromagnetism, and thus, he hoped, explaining the existence of all particles and interactions in nature. Could the properties of electrons, for instance, be deduced though general relativistic theory? Before he could flesh out his ideas on the subject, he received word from the mathematician Hermann Weyl, who claimed to have found his own unified approach.

Weyl, a stellar figure in mathematics and physics, was educated in Göttingen under Hilbert and moved on to a position at the ETH, where his career briefly overlapped with Einstein's. (Their paths would merge again, much later, when they both worked at Princeton.) He took special interest in the geometric basis of the nascent general theory of relativity, and in 1918 wrote a pivotal exposition of the subject, entitled *Space, Time, Matter.* He sent the printer's proofs to Einstein for his inspection before publication. At the same time, Weyl informed Einstein, "Lately I have succeeded in deriving electricity and gravitation from a single common source,"[17] and asked if he could send him a copy of the manuscript describing this feat. He requested that Einstein, in his capacity as director of the Prussian Academy, submit the paper for publication in its *Proceedings.*

A week later, Einstein wrote back, glowing with enthusiasm about the book, calling it a "symphonic masterpiece."[18] About the unification proposal, Einstein was most excited. "You have given birth to the child I absolutely could not muster: the construction of Maxwell's equations out of the [metric tensor coefficients]. . . . Naturally, I am tremendously eager to see you and your paper."[19]

Weyl sent Einstein a copy of his article, entitled "Gravitation and Electricity," which attempted to unify those two forces by use of a modified four-dimensional space-time geometry. Weyl multiplied the terms in the metric of general relativity by a new factor called a *gauge,* which is an arbitrary function of the coordinates.

The gauge comes into play in electromagnetic theory in relationship to something called the electromagnetic vector potential. Briefly, the electromagnetic vector potential is a four-vector (directed arrow) from which the electromagnetic field tensor derives. In other words, knowing one gives you the other. But physicists have long known that adding the differential (a calculus operation) of any function one can think of to the vector potential still produces the same field tensor. The arbitrary nature of this function reminded researchers of changing thickness, pressure, rainfall, and other gauged quantities, hence the term *gauge.*

This definition is rather technical, so here is an analogy. Imagine that there is a new national lottery. If you select the correct twenty-digit lottery number, you are the winner. To sign up and receive your ticket, you must call one of the special phone numbers designated by the lottery agency. To make things easier, the agency has arranged that its phone numbers can be called with any area code in the country. No matter what area code one chooses, one still gets through and obtains the same ticket. By analogy, the lottery number is like the electromagnetic field tensor, a detailed set of information. The phone number is like the vector potential, a way of accessing this information. Finally, the gauge is like the area code, an appendage to the phone number that in this case makes absolutely no difference.

Deciding that what's good for the goose is good for the gander, Weyl inserted a gauge term into the metric of general relativity. What the extra factor buys is a new set of equations, in addition to those modeling gravitation, that can be adjusted to reproduce Maxwell's equations of electromagnetism. However, the cost is dear. Weyl's non-Riemannian theory implies that lengths, times, and even the Gibraltar of relativity, the hitherto invariant space-time interval, can be stretched or compressed by a changeable amount.

In Weyl's approach, a four-vector transported parallel to itself along two different paths might not only end up pointing in different directions, but also having different lengths. In our twins example, imagine Marius and Darius setting out in different directions, reuniting, and then finding out that one has become a foot shorter than the other. The same variation would take place in the readings of any clocks they were carrying. Moreover, unlike in special relativity, where different observers can have different readings, these changes would take place according to all observers; they would be absolute effects.

Einstein found these variations in length and time scales troublesome and unphysical. On the other hand, he was impressed by the diligence of Weyl's effort and felt that the paper could be published with an addendum stating how it diverges from physical reality. In a letter expressing these feelings, Einstein comes both to bury Weyl and to praise him. "Except for the agreement with reality, it is in any case a grand intellectual achievement."[20]

Weyl was grateful for Einstein's willingness to publish his paper, but he was frustrated by his inability to convince Einstein that his theory was correct. "Your rejection of the theory weighs heavily on me," he wrote Einstein. "I know only too well how much closer a contact you have with reality than I. But my own brain still keeps faith in it. . . . If in the end you are right, then I would regret having to accuse God almighty of a mathematical inconsequence."[21]

Einstein and Weyl went back and forth, arguing whether it made sense to rewrite relativity in such a manner. Neither could convince the other. Then a paper shortage prevented Einstein from submitting Weyl's article; it ran over the limit of eight pages. Einstein wrote to Weyl, explaining this fact and telling him, "I have studied your paper but am more than ever convinced that you have gotten onto a very dubious track which is regrettably costing you your valuable energy."[22]

Weyl was extremely disappointed that someone he so deeply respected disagreed with him so vehemently. "I am caught between faith in your authority and my view," he wrote to Einstein in a letter once again defending his work. He ended his letter quite sharply. "Even though the war hatchet between us has been dug up, I give my regards in sincere respect."[23]

Weyl's unified field theory never caught on, because of its manifest physical problems. His gauge approach in general, though, proved an enormously powerful tool for modern physics, particularly for quantum models. Sometimes a key unlocks different doors than expected.

A Letter from Königsberg

Just when Einstein thought he was free of grappling with one mathematician's unified field theory, he needed to wrestle with another's idea. In April 1919, he received a letter from Theodor Kaluza, an unknown lecturer at the University of Königsberg. Kaluza requested that Einstein review his notion for the unification of gravitational and electromagnetism in five dimensions. Unlike Nordström, Kaluza based his work on the fully developed general theory of relativity. Furthermore, in contrast to Weyl, Kaluza left the basic principles of general relativity intact, enhancing rather than changing it by adding an extra dimension. Remarkably, with this extension, he found that he could reproduce both Einstein's gravitational theory and Maxwell's equations of electromagnetism with a single set of relationships.

Einstein was intrigued. Somehow this obscure lecturer from a remote part of east Prussia had seemingly achieved what the great professors of the major universities could not. Yet, then again, that's how Einstein had started out, when he set aside the logbooks of a

Swiss patent office and logged his name in the annals of history. Keenly aware of his past, he was inclined to give the unknown researcher a chance to prove himself.

But Einstein knew that the idea of a detectable fifth dimension was completely unphysical. As Ehrenfest had pointed out, many aspects of nature would appear quite differently if the world had more than three spatial dimensions plus time. Einstein was well aware of the discomfort many scientists felt when contemplating time as the fourth dimension. What then would they think of five dimensions? Would they see it as the realm of science or the world of ghosts?

After sending Kaluza some questions about his work, soon Einstein had other vital matters to address. In November 1919, two British teams reported spectacular results concerning one of the key predictions of general relativity. They measured the bending of starlight by the Sun during a solar eclipse, and found that it matched Einstein's estimates. The findings were paraded in newspapers all over the world. Einstein was no longer just a prestigious scientist in Germany. He was the most famous physicist in the world. Given his unmatched influence, how should he respond to Kaluza's promising but strange new theory?

CHAPTER 5

Striking the Fifth Chord
Kaluza's Remarkable Discovery

In that blessed region of Four Dimensions, shall we linger on the threshold of the Fifth, and not enter therein? Ah, no! Let us rather resolve that our ambition shall soar with our corporal ascent.

—EDWIN ABBOTT, *Flatland*

Bedtime Story

"Papa, tell me a story," the little girl asked.

Theodor Kaluza's mind was full of worries. As a privatdozent of mathematics, he lived from day to day, literally dependent on how many places he could fill in his classes. Student by student, the pennies trickled into his busker's cap. How could he ever get a permanent job if he didn't publish more? But he was a dreamer much more than a doer, and preferred quiet reflection about ideas more than writing them down in journal articles. He only published what he deemed his worthiest results—a paper every few years on average. Sitting in his study, reading a good book, tinkering with geometric models, and taking breaks to spend time with his family was much more satisfying than the academic rat race. Yet he needed a steady salary somehow.

"Papa, please, I'm waiting."

Kaluza enjoyed playing with his children and encouraging their intellectual interests. Like Ehrenfest, he wanted his kids to have a broad education with a strong mathematical component. Rather than overwhelm them with pedagogical material, however, he had a much gentler approach. He thought of just the bedtime tale to tell his daughter: the story of Flatland.

As his daughter Dorothea lay smiling in bed, he proceeded to tell her about a flat kingdom full of creatures that know nothing of a greater world. Scampering about, they live their lives in ignorance of anything beyond their thin sheet. These bedbugs . . ., he mused.

Bedbugs? To enliven his version of Flatland and elucidate its two-dimensional world, Kaluza made the leading characters bedbugs.[1] Living within limited means, these insects may have been quite familiar to him and his family. Under such circumstances the expression "good night, sleep tight, and don't let the bedbugs bite," would represent a real concern. Or perhaps in his retelling, he was referring to Gauss and Sylvester's bookworms or Helmholtz's two-dimensional creatures in their earlier (pre-Abbott) versions of the tale.

Kaluza biographer Daniela Wünsch speculates that Theodor Kaluza may have learned about Flatland through the profession of his father, Max Kaluza.[2] As a specialist in English language and literature, Max Kaluza had a very similar background to Edwin Abbott and was likely familiar with his work. Thus, conceivably the story of Flatland could have been one of Max's bedtime tales to Theodor when he was a boy as well.

An English Childhood in Prussia

Theodor Franz Eduard Kaluza was born in the German town of Wilhelmsthal on November 9, 1885. Coincidentally, fellow unifier Hermann Weyl arrived in the world the exact same day. Kaluza's father was from the nearby town of Ratibor, where he could trace his lineage back three centuries. His mother, Amalie, of aristocratic heritage, was from a village near Ratibor as well. Both Wilhelmsthal and Ratibor are situated in the Prussian province of Upper Silesia, now part of Poland.

The Kaluzas had a long-standing tradition of education and culture. Each generation produced at least one teacher, priest, or important government official.[3] One distant ancestor dutifully worked as a provincial inspector for Frederick the Great. Another, Augustin Kaluza, was a Catholic theologian who studied the natural history of Silesia. He published several notable treatises on the minerals and wildlife of the region.

Max Kaluza's scholarly contributions brought him well outside of Silesian life. Though a native German, his heart lay in merry old England, where it embraced the ancient tales of Chaucer. He edited Chaucer's *Romaunt of the Rose,* a task that required mastery of fourteenth-century Middle English. Familiar with Old English as well, he wrote several books on the history of English versification and grammar. High regard for his scholarship landed him a position at the University of Königsberg, where he moved the family two years after Theodor's birth.

For young Theodor, Königsberg was a wondrous place, full of music and life. It was a great regional center of culture, bursting with theaters, orchestra halls, and opera houses. In the intellectual arena, not only was it famous as the home of Kant, it was also known for its mathematical tradition, dating back to Euler. Virtually every mathematics student learns at some point about Euler's Königsberg bridge problem, a puzzle based upon the city's intricately connected riverbanks and islands. Seven bridges cross the river Preger, and Euler proved that it was impossible to walk across them all without traversing the same one twice.

The bridges of London, on the other hand, could be crossed in sequence without repetition. Thanks to his father's Chaucerian pursuits, Kaluza had the opportunity to see them himself. It was rather exciting for Kaluza to travel with his family to England during research visits. One memorable tour included a Shakespearian jaunt to Stratford upon Avon. With the swiftness of the Artful Dodger, Kaluza mastered English, and was reading Dickens by the age of nine. In addition to standard English, he learned how to speak various dialects and slang. Another summer journey to Hungary brought him knowledge of Hungarian. Eventually he would become familiar with seventeen languages, including Lithuanian, Arabic, and Hebrew. Of these, he spoke seven with perfect fluency.[4]

In general Kaluza's childhood was a happy one. A friendly, inquisitive lad, he found great joy in intellectual discovery. Math came easy to him, but he was also adept at many other subjects. He felt that the mind could overcome almost any hurdle. He demonstrated this to himself, when, later in life, he read a book about swimming and then swam successfully on the very first attempt.[5]

Life's Essentials

Kaluza often liked to say, "In our civilization the most important decisions are selecting one's marriage partner and choosing one's career."[6] He wasted no time attending to both choices. In 1903, he began his studies in mathematics, physics, and astronomy at the University of Königsberg. Under the supervision of F. W. F. Mayer, he conducted his thesis work on a mathematical technique called Tschirnhaus transformations. In 1906, while his brain was busy investigating this research, his heart first encountered Anna Beyer, a businessman's daughter and the sister of one of his classmates.[7] They would marry three years later, rounding out the other side of his life.

To broaden his mathematical experience, in 1908, Kaluza decided to spend a year at the University of Göttingen. It was an extraordinary interlude in the young thinker's life. At that time Göttingen's math department was truly at its peak. The legendary Felix Klein, though retired from active research, remained an influential teacher and editor. His venerable presence attracted talent from all over the world. David Hilbert, his protégé, had completed a groundbreaking analysis of the fundamentals of geometry and was developing the mathematical underpinnings of modern physics. His theory of infinite-dimensional Hilbert spaces would become the basis of quantum mechanics. Last but not least, Hermann Minkowski was championing his revolutionary concept of space-time and his critical reinterpretation of special relativity. Who could match such a dynamic trio?

Not to mention the prodigious students and visiting researchers strolling the town's medieval streets at any given time. The spirits of Gauss and Riemann infused many of them with the bold desire for unity. For those hoping to compose the universal symphony, Göttingen was none other than the Juiliard of mathematics.

During the years preceding Kaluza's visit, its streets were filled with prospective unifiers. A cafe owner on Weender Strasse near the marketplace could well have seen Paul and Tatyana Ehrenfest walking by, pushing their baby daughter in a carriage, chatting about an interesting lecture by Klein or Hilbert On another day, Gunnar Nordström might have been making pace toward the auditorium, anxious to hear the resounding words of Minkowski. Past the same windows on a different occasion might have sauntered Hermann Weyl, deep in thought, contemplating possible solutions to a tricky set of equations. Each of these Göttingen dwellers ended up making vital contributions to ideas about unified field theories and/or higher dimensions.

When Kaluza arrived, he became acquainted with Weyl.[8] They undoubtedly went to many of the same talks. Ehrenfest and Nordström had already left, so Kaluza missed getting to know them too. For each of these thinkers, it's clear that their common experiences in Göttingen helped shape their future research. By being at the epicenter of Minkowski's thunderous four-dimensional fusion—the reverberating consummation of Einsteinian relativity—ultimately each would be inspired to conduct his own unification experiments as well, with varied degrees of success.

In 1909, Kaluza returned to Königsberg with a renewed sense of mission. Like a potter with a fresh supply of clay, he awaited the opportunity to mold the nascent relativistic formalism. But first he needed permission to teach at the university. This was granted when he passed his habilitation exams.

Only twenty-four when he began, the enthusiastic new instructor was dark-haired and handsome, looking even younger than his age. He could easily pass for one of the students. This was a problem for the strict administrators, who viewed the divide between lecturers and students as an unbreachable frontier. They quickly ordered him to cease and desist—from shaving that is. After hearing the news directly from the dean, Kaluza—always the mathematician—gleefully calculated how many square meters less shaving he would need to do in the future.

He grew a thick black beard which, although he was Catholic, gave him a rabbinical appearance. This led to a certain amount of teasing. One time an impertinent girl called out to her friends while he was walking by, "Oh look, truly the genuine item from Palestine!"[9] Much later, when teasing became life-threatening harassment under the Nazis, Kaluza decided to shave it off, going beardless from 1933 onward.

Kaluza was known as a friendly but not particularly dynamic teacher. Mathematician Martin Kneser, who took one of Kaluza's courses (at Göttingen in the 1940s) recalls, "Although he was a kind man, I don't think he was a very inspiring teacher. His course, as far as I can remember, was nothing extraordinary. It was not really a good course."[10]

One unusual aspect of his teaching is that he lectured completely from memory. He reportedly referred to his notes only once in his career: to copy a fifty-digit number onto the board.[11]

Kaluza also had a reputation for being absent-minded—a recognized hazard of his occupation. This is illustrated by the following story. One evening Kaluza failed to show up for his class in number theory. The students waited and waited, but still no instructor. Finally some of them gave up and decided to head over to a new

Theodor Kaluza, the German founder of the five-dimensional extension of general relativity, photographed in his later years.

Chaplin film. The movie had already started, but they wanted to sit near the front of the theater. As they made their way forward, they stumbled across an older gentleman. It was Kaluza. He had seen the advertisement for the movie, thought it looked interesting, and completely forgotten that he had to teach that evening.[12]

In 1910, Theodor Jr., the Kaluzas' first child, was born. Kaluza liked to joke that his son cost him only one mark (a fraction of a dollar). The reason was that the university hospital customarily treated faculty members for free. When Kaluza wouldn't hear of that, they sent him a bill for the paltry sum of one mark—to be paid cash on delivery.[13]

Dorothea, the Kaluzas' second child, arrived six years later. There's no record of how much she cost. Nevertheless, she had to pay an awful price herself. During the first two years of her life, her father was called to the Western Front, to battle in World War I.[14] Anna had to raise the children on her own until Theodor safely returned in 1918.

A Secret Crescendo

After the war, Kaluza resumed teaching at the university and doing mathematical research at home. By then he had been at the level of privatdozent for a decade, and greatly needed more widely respected publications to move forward. He had published several articles already, including an important paper on the problem of a relativistic rotating disk, but was little known in the academic community.

Nevertheless, his primary goal was to become reacquainted with his family. While bright young Theodor Jr. idolized him, Dorothea hardly knew him, and he hoped to spend ample time with both. His son received an open invitation to sit in his study and watch him whenever he was working.

By that time, Kaluza had become familiar with Einstein's equations of general relativity. Playing around with this formalism like a cat with a new toy, he decided to see what would happen if he added an extra dimension. Suddenly he came to a startling revelation: by extending Einstein's gravitational equations he could reproduce

Maxwell's theory as well. The fifth dimension, he realized, would allow room for a unified theory of nature! Kaluza's son later described that moment of inspiration: "He sat completely still for several seconds, and then he whispered very sharply and banged the table, and he stood up but remained completely motionless for several seconds. Then he began to hum the last part of an Aria of Figaro."[15]

Realizing the importance of his results, Kaluza encapsulated them in an article, "On the Unity Problem of Physics," and sent it off to Einstein. Beginning with the fundamentals of space and time, he detailed his scheme for unification step by step. He hoped to convince Einstein, and the readership of the journal Einstein edited, that by grafting an extra dimension onto the tree of general relativity, it could produce unexpected fruit. Aware of Weyl's foray into this arena, Kaluza strived to do even better.

> A few years ago H. Weyl made a surprisingly bold thrust toward the solution of [the unification] problem, one of the great favorite ideas of the human spirit. Disregarding the difficulties which accomplish H. Weyl's deep-seated theory, ideally, one can imagine an even more perfect realization of the quest for unity: the gravitational and electromagnetic fields stem from a single universal tensor. I wish to show here, that such a tight union of both world powers seems possible in principle.[16]

Kaluza began his exposition by extending the metric tensor (the mathematical rule sheet used in general relativity to determine the space-time intervals between events) by adding extra components pertaining to a fifth dimension. If one thinks of the four-dimensional metric tensor as a four-by-four checkerboard, with each row and column corresponding to a dimension (three for space and one for time), then Kaluza simply added one extra row and one extra column. The revised tensor possesses as many slots as a five-by-five bingo board, housing twenty-five separate mathematical expressions. This is nine more than usual, but not all of these are independent. Four of the extra ones must be identical to another four, plus there is one more. This yields five independent new components.

Kaluza equated four of the five new metric constituents with the four components of the electromagnetic vector potential. Recall that this is the "phone number" with which one can "dial up" the electromagnetic field tensor, the principle actor in Minkowski's rendition of Maxwell's equations. Not knowing what to do with the fifth new metric component, Kaluza designated it as a scalar (later called the dilaton or Brans-Dicke scalar). He also took the standard stress-energy tensor, describing matter and energy, and framed it with the components of the electromagnetic four-current. Once again, this is like adding new squares to a checkerboard, then filling them with new pieces.

Then, following Einstein's procedure, Kaluza calculated the connections associated with the complete metric, including the new terms. In standard general relativity, the connections sew together the space-time manifold, producing the warping effects of gravitation. However, this super-sized version yields an abundance of additional connection terms that (with some extra assumptions) miraculously resemble the components of the electromagnetic field tensor. Next, Kaluza computed the curvature tensor, a function of all the connection components. This similarly produces standard terms as well as new elements.

Eager to test out his new recipe, Kaluza processed everything through Einstein's equations of general relativity and sampled the end product. He was left with two concoctions. One was the exact form of the standard equations for gravity; the other, precisely Maxwell's equations for electromagnetism. His recipe worked; one preparation yielded the two then-known fundamental forces of nature. It was just perfect to serve to the physics community to satisfy its appetite for simplicity.

"Voila!" the great equation chef said, lifting up the cover. "In addition to your fabulous main course, gravitation au gratin, you get a mouthwatering dessert, electromagnetism à la mode, for free."

But not all of this is savory, as Kaluza conceded. He admitted in his paper that "all our previous physical experience hardly provides any suggestion of [a fifth dimension]." To address such concerns, he cleverly included in his work a hypothesis called the cylinder condition. It mandates that none of the physically measurable elements of the theory can depend on the fifth dimension. Therefore the fifth dimension is undetectable and its presence can't cause problems. The term *cylinder condition* stems from Kaluza's envisioning of the fifth dimension as cyclical rather than linear. Any motion in this direction leads right back to the beginning and is therefore as imperceptible as the whirling of a high-speed propeller. The only movement that counts takes place in the other four space-time dimensions, preserving the observed characteristics of nature. How did he justify this premise? He didn't provide any physical basis for it, but rather asserted its necessity to counteract "the strongly alienating decision of calling the fifth dimension to the rescue."[17]

Einstein's first response to Kaluza's theory was wholly positive. "The idea that [unification] can be achieved by a five dimensional cylinder-world has never occurred to me and would seem to be altogether new. I like your idea at first sight very much. From a physical point of view it appears to me far more promising than the mathematically so penetrating Ansatz of Weyl."[18]

But then, as Einstein started to work through Kaluza's equations in more detail, he began to see some issues. One week after his first letter to Kaluza, he wrote again, "I have read through your paper and find it really interesting. Nowhere, so far, can I see an impossibility. But on the whole I have to admit that the arguments brought forward so far do not appear convincing enough."[19]

Einstein did not believe Kaluza had thought through the physical consequences of his theory well enough. Einstein asked him, in particular, the critical question of what would happen to a charged particle under the simultaneous influence of both gravitational and electric fields. Would it act in the manner predicted by experiment or behave completely differently? He also warned Kaluza that his paper was too long (like Weyl's earlier effort, it had surpassed the sacred page limit), and suggested that he try another journal. Kindly, however, Einstein left open the possibility of publication in *Proceedings,* if Kaluza addressed these issues and couldn't get his article placed elsewhere.

Two years passed. Reconsidering Kaluza's work, which had never been published, Einstein realized that it would be a loss if the scientific community missed the chance to read it. Therefore, he wrote to Kaluza, "I am having second thoughts about having restrained you from publishing your idea on a unification of gravitation and electricity two years ago. Your approach seems in any case to have more to it than the one from H. Weyl. If you wish I shall present your paper to the Academy after all provided you send it to me."[20]

Kaluza immediately sent Einstein a fresh copy of his article. The revised version addresses some of the issues Einstein brought up, but concludes that "even in the face of all the physical and epistemological difficulties which we have seen piling up against the conception presented here, it is still hard to believe that all these relations in their virtually unsurpassed formal unity, should amount to the alluring play of a capricious accident."[21] Einstein sent the paper to the academy, which published it in its *Proceedings* in December 1921.

In the next few years, Kaluza's paper languished in the same obscurity as its author. In 1922, Einstein, along with mathematician Jakob Grommer, wrote a minor article about the subject, showing that Kaluza's theory, leaving out the stress-energy tensor, cannot describe the behavior of an electron without containing mathematical monstrosities called singularities. Singularities often form the stuff of physicists' nightmares, because they can replace workable calculations with irremovable infinities.

Leaving out the stress-energy tensor was important to Einstein, because it explicitly contained the matter terms. Like Clifford, Einstein believed that a proper unified theory would produce matter through geometric properties, rather than by specifically including it.

The same year Einstein wrote a letter to Weyl, asking, "Have you thought through Kaluza's attempt? Although it initially strikes me as closest to reality, it also fails to provide the singularity-free electron. I believe, in order really to progress, we must once again find a general, fundamental principle of Nature."[22]

For the next few years Einstein worked sporadically on the question of unified field theories, mainly publishing comments about or generalizations of the work of others. His papers during that period included a few brief works regarding an approach by English physicist Sir Arthur Eddington and one theory of his own somewhat connected to Weyl and Eddington's ideas. Aside from the Grommer collaboration, none of Einstein's work at that time addressed the fifth dimension. Of his homemade model, which he soon abandoned, he wrote to Ehrenfest, "I have once again a theory of gravitation-electricity; very beautiful but doubtful."[23]

The Mathematical Cinderella

Meanwhile, the Kaluzas found themselves in increasingly dire straits. Inflation ate away at their meager income, forcing them to tighten their belts more and more. Not that they were starving; rather they had to make due with the bare necessities of life. As someone who grew up under much better circumstances, Anna Kaluza often felt overwhelmed by their distressing situation. Kaluza's son recalled his mother standing by the cupboard and crying, gesturing how little food was left. Yet somehow they always had something to eat, even if they had virtually no money left over for toys or other luxuries.

Even in the worst of it, Theodor Kaluza always kept his magnanimity. Once his son, similarly gifted in school, won a special scholarship. Kaluza knew that if he accepted the funds, a certain poor widow's child would lose them and be unable to attend school. Without hesitation, he declined the scholarship, winning tremendous gratitude from the school's headmaster.

As his need for a steadier income grew ever more pressing, in 1925 Kaluza set his humility aside and decided to write to Einstein about his circumstances. He indicated in the letter why hope for a professorship forced him to abandon further efforts at physical unity and focus instead on publishing more mundane mathematics articles. As he informed Einstein: "I can devote myself now to only a little physics, because my mathematical work occupies me too strongly, particularly because I must strive ultimately to become better known through more intensive publishing—and thus perhaps preparing an end to my local, unsatisfactory, Cinderella-like existence."[24]

Einstein was greatly moved by the poor instructor's plea. Expressing the hope that a steadier position would allow him the time to resume his promising efforts toward unification, Einstein replied, "I am still of the opinion that your idea to construct a relation between electricity and gravitation is of great originality and merits the serious interest of academic colleagues. Besides the Weyl-Eddington idea it is the only attempt to be taken seriously in that direction. It would be desirable that you soon find time and leisure to tackle those problems again. I myself have so far struggled with this problem in vain."[25]

Over the next couple of years, Einstein contacted other professors in attempts to secure an academic position for Kaluza. In one such letter, he wrote glowingly to the Austrian émigré physicist Karl Herzfeld of Johns Hopkins: "The internal relationship between gravitation and electricity I believe to have found now after long erring travels, in close connection to an idea of Kaluza, which appeared some years ago (1921) in our academy reports. One should try to provide for him favorable possibilities for employment."[26]

Alas, Herzfeld couldn't hire Kaluza. He had just joined the Hopkins faculty himself and was not in the best position to promote an unknown instructor. Finally, Einstein learned about an opening at the University of Kiel. The director there was Adolf Fraenkel, a mathematician specializing in set theory. Trusting Einstein's judgment, he gladly appointed Kaluza to be professor at Kiel. At last, the poor privatdozent had a steady job. Cinderella's prince had come to the rescue.

While looking for positions for one five-dimensional explorer, Einstein found out about another venturer into similar terrain. In June 1926, Paul Ehrenfest sent Einstein one of his irresistible invitations to come to Leiden. In trying to lure visitors to his celebrations of the intellect, Ehrenfest usually provided a list of who else was there already, as if to say "come and join the party." This time, Oskar Klein was the featured guest. Ehrenfest, in typical style, extolled Klein's virtues, mentioning that he was Niels Bohr's favorite young scholar. Einstein declined, however, because of family matters.

Nevertheless, after finding out from Grommer about Klein's studies of five-dimensional models, Einstein became intrigued. He wrote to Ehrenfest asking for a copy of a recent article by Klein. Klein sent it to Einstein himself, expressing his joy that the founder of relativity would be interested in his work. In September 1926, Einstein read over Klein's paper and wrote back to Ehrenfest that it was "beautiful and impressive."[27] He noted, however, his misgivings about Kaluza's cylinder condition, calling it unnatural. Inspired in part by Klein's findings but eager to reestablish the theory on firmer ground, Einstein would soon resume his own studies of higher-dimensional unification.

CHAPTER 6

Klein's Quantum Odyssey

A main problem for us theoreticians rather resembles that represented by Charybdis and Scylla, between which Odysseus was forced to steer. . . . Speculation is certainly a necessary part of theoretical work, just as much as building on experimental facts. Still, it drags many of us into a mental whirlpool not unlike the hydrodynamical one of Charybdis, from which the escape feels like a miracle. On the other hand, sticking too closely to facts—Scylla had six hard ones—may be equally deadly when using them as building stones for a theory.

—OSKAR KLEIN, From *My Life of Physics*

The Mariner's Journey

Oskar Klein was a scientist of striking contradictions. An international traveler with a global perspective, he spent much of his latter career at a relatively isolated institute in Sweden. A lover of the practical, experimental side of science, he nevertheless relished philosophical discussion of the most abstract sort and pressed far-flung theories to their conceptual limits. A serious hard worker, he was blessed with a fabulous gift of satire. One of the leading proponents of the fifth dimension, he once drank to its death.

In an autobiographical essay, Klein compared his own struggles to those of Odysseus. Like the legendary Greek hero, he saw himself as a mariner steering a course between the two deadly extremes of stodgy practicality and whimsical imagination. Coming and going from Stockholm, his own personal Ithaka, Klein relished exploring distant lands, but also looked forward to the familiarity of home. He enjoyed his encounters with mysterious creatures of the deep— strange new physical effects—as much as he appreciated the quiet everyday shipboard tasks of a scientist.

Though a highly original thinker, Klein is best known for his "alsos" and "almosts": In quantum physics, he is cohonored with Walter Gordon for the Klein-Gordon equation of relativistic waves, with Pascual Jordan for the Jordan-Klein matrices and second quantization theory, and with Yoshio Nishina for the Klein-Nishina formula of high-energy photon scattering by electrons—not to mention his sequel to Kaluza. Of his most famous results, only the Klein paradox, an effect related to the reflection of electrons by a barrier, comes with no dash before or after his name.

And the "what ifs" are legendary. Klein reportedly developed an early version of the Schrödinger equation (the dynamo of quantum mechanics), but was too ill at the time to see it through to publication, and proposed a description of the strong interaction that featured premonitions of Yang and Mill's pivotal gauge theory. And when Kaluza-Klein theory made its late twentieth-century comeback virtually at the time Klein had just left this planet, one could almost hear his spirit cry out, "I told you so!"

This record of collaborative hits and tantalizing near-misses does not detract from the value of Klein's contributions. On the contrary, it points to his pivotal place at the heart of twentieth-century physics. Like Werner Heisenberg, Erwin Schrödinger, Paul Dirac, and other notables, Klein seized on the black box of atomic behavior and never relented in his quest to explain its mysterious inner workings. His innovative outlook inspired many others to set aside their preconceptions and ponder alternative explanations.

For much of his life, Klein lived in the shadow of the Nobel Prize. This was only natural, since he was a member of the small community of physicists in the city where the prize was awarded. The connections were even more than his nationality would suggest. His first research, at the Nobel Institute, was under the guidance of Nobel laureate Svante Arrhenius. For many years, Klein was one of the leading assistants to Niels Bohr, another recipient of the prize. He was also close friends with Wolfgang Pauli and had many contacts with

Heisenberg, Schrödinger, Dirac, Max Born, Hideki Yukawa, and, later in his life, Abdus Salam, all laureates (or laureate-to-be, in Salam's case). Klein served on the Nobel committee but never received its blessing, even though the collective importance of his scientific and humanitarian achievements clearly could have warranted him such a distinction. If Nobel issued a lifetime achievement award, it would have fit him well. And if there was a "noble prize" for pure human decency, he would have won it hands down.

It takes no Livingstone or Stanley to discover the source of Klein's ever-flowing generosity and zeal for knowledge. His father, Gottlieb Klein, the first chief rabbi of Sweden, was a sage of great wisdom, a lover of learning, and a well-respected unifier of faiths. Born in the Slovakian town of Humenneh, nestled in the Carpathian mountains, Gottlieb Klein grew up in a region similar to the one from which Kaluza's family derived. Leaving home at a young age, he wandered through central Europe, traveling to Heidelberg, where he received a doctorate. Exploring a full spectrum of intellectual pursuits, he attended lectures by famous scientists such as Helmholtz, Bunsen, and Kirchhoff. After completing rabbinical studies, he moved to Stockholm to lead the fledgling Jewish community of mainly German expatriates. There, he married Antonie Levy, the daughter of a German scholar of Eastern studies.

Though he could speak no Swedish at first, Rabbi Klein soon assumed a vital place in the Swedish establishment. He became close friends with the king of Sweden, who was very interested in the philosophy of religion. His liberal views and advocacy of interfaith dialogue won him the friendship and respect of the high-ranking clergy in the established church. He also came to know Sweden's leading thinkers, including Arrhenius. Comfortable with secular pursuits, it didn't seem to bother him that none of his own children grew up to be religious. His greatest wish and blessing was that they would conduct meaningful humanitarian lives—a gift for which they were forever grateful.[1]

Youthful Chemistry

Born in Mörby, a Stockholm suburb, on September 15, 1894, Oskar Benjamin Klein, the rabbi's youngest son, wanted to be a scientist almost from the start. An avid collector of shells, butterflies, and other specimens of nature, he loved watching the stars through his mother's opera glasses. With childhood readings that included the works of Darwin, his earliest ambition was to be a biologist. But when, as a teenager, his interests turned to chemistry, his parents bought him an entire laboratory—not just a basic set, but a fully stocked workroom. To guide him in his research, they gave him an

Oskar Klein, one of the foremost Swedish physicists of the twentieth century and the developer of five-dimensional theory in its quantum form.

instructional book by the chemist Wilhelm Ostwald. Klein read the book with great enthusiasm, committing it almost to heart. Soon he was performing sophisticated experiments with various materials, including making his own fireworks.

One day in the summer of 1910, Klein's father was invited to a peace conference. Finding out that Ostwald would also be there, he eagerly took up the offer. After the conference, Arrhenius welcomed Ostwald and the rabbi to come over for lunch. The rabbi asked if he could bring two of his boys along: Klein and his brother. Arrhenius agreed, and was delighted to meet children so fascinated by chemistry.

Impressed by Klein's youthful interests, Arrhenius invited him to join his research lab. He assisted Arrhenius on a number of projects related to radiochemistry, using a primitive electroscope to examine radioactive decay products. In his free time after school, he also collaborated with Arrhenius's deputy, Ernst Riesenfeld, on a research article about the solubility of zinc hydroxide in alkalis. Thus by age eighteen, Klein was already a published author and full-fledged member of the experimental community.

Around that time, he finished high school and began attending Stockholm University. He continued to study chemistry, but also took up physics after reading texts by Lorentz and Helmholtz. Surprisingly, Arrhenius suggested that theoretical physics might very well be the best career for him. Klein delved into his newfound topic with great anticipation. "These were wonderful times," he later recalled, "when all was new and my eagerness almost unlimited."[2]

In 1914, Klein's father died and his mother thought it would be a good idea for the grieving youth to spend some time abroad. He embarked on research trips to Germany and France, during which time World War I broke out. He tried to remain in Paris to work with experimental physicist Jean-Baptiste Perrin, but found himself recalled back to Stockholm. From June 1915 until October 1916, he completed his compulsory military service. Though an internationalist at heart, he was also patriotic and felt proud to serve in the Swedish army.[3]

The Riddle of the Atom

Returning to Arrhenius's lab after his discharge, Klein began to study for his Licentiat, the exam certifying the completion of his training. He started to read more theoretical physics, including the papers of Bohr. This was his first real exposure to the idea of quantization: the principle that energy comes in discrete packets.

Bohr's model of the atom explains why the light produced or absorbed by gases breaks up, under analysis, into fixed spectral lines. It details the radial positions of electrons as they orbit the nucleus of an atom and mandates that they can jump from place to place only by releasing or consuming exact units of energy, called quanta. Whenever an electron drops closer to the nucleus, it gives off a light quantum, or photon. Whenever it moves farther away, it takes in a photon. Otherwise, it can maintain a stable orbit for an indefinite period of time. The frequencies (rates of oscillation) of the emitted or absorbed photons are determined by their energies. Because each frequency manifests itself as a distinct spectral line, the energy levels of atoms display themselves in characteristic spectral patterns—unique rainbows revealing their inner workings.

The basis of Bohr's theory was difficult for Klein to understand at first because of Klein's practical background. Then, an emissary from Bohr's laboratory, the Dutch physicist Hendrik Kramers, came to Stockholm to give some lectures. Klein was impressed and felt that Kramers, though roughly his own age, appeared to be much older and more experienced.

Around that time, Klein found out about a research fellowship. Although he originally considered requesting funds to work with Einstein or Peter Debye (another physicist), he was so impressed by Kramers that he decided to apply for a position with Bohr instead. Granted the funds, he journeyed down to Copenhagen, where Bohr worked in a small office as Denmark's first professor of theoretical physics.

Klein's decision portended major changes that were taking place in the world of physics. During Bohr's remarkable tenure, the physics community's center of gravity would move north from Berlin, the capital of Bismarck, the Brandenburg Gate, and Prussian imperial order, to Copenhagen, the home of Hans Christian Anderson, the Little Mermaid, and boy soldiers marching in the gardens of Tivoli amusement park. In this magical northern location, miraculous new theories would soon be born.

Once in Copenhagen, Kramers, who was Bohr's chief assistant, took Klein under his wing and gave him a crash course in theoretical atomic physics. Klein soon appreciated the pressing problem of the times: explaining the underlying reasons for the electron's strange behavior.

At that point, Bohr's atomic model was in serious trouble. Although it successfully predicted some of the spectral line patterns, it couldn't explain why the electrons circling a nucleus stay in their orbits for a time, then jump, by discrete amounts, down to lower energy states. In other words, it captured some of the rhyme, but couldn't supply the reason. However, Bohr was hopeful that science would eventually be able to interpret this unusual behavior, possibly leading to a new way of thinking about physical reality.

While on a walk in the countryside north of the city, Klein came to appreciate Bohr's unique philosophical perspective. Similar to Taoists, Bohr saw nature as the union of opposites. This attitude later led him to embrace the principle of complementarity. Bohr's complex mind readily accepted sharply contradictory views, treating them like interlocking pieces of the same puzzle. For example, though in some ways (by defining its energy and momentum) he treated the atom as a mechanical system, in other respects (by mandating discrete orbits) he eschewed the laws of mechanics altogether. As he once remarked, "We will never understand anything until we have found some contradictions."[4]

During 1918 and 1919, Klein traveled back and forth between Copenhagen and Stockholm several times, engaging in research projects in the laboratories of both Arrhenius and Bohr. Meanwhile, after Kramers returned to Leiden to complete his doctoral studies with Ehrenfest, Bohr appointed Klein to be his principal assistant. Thus began a long period of fruitful collaboration between Bohr and Klein.

Bohr was extremely athletic, famous in Denmark for his exploits on the soccer field as much as for his scientific prowess. He loved hiking, sailing, and skiing, and took every opportunity to exercise his body while engaging his mind in deep thought about quantum matters. He and Klein, who was lanky but also in good shape, took turns hosting each other on various such excursions, which gave them plenty of time to share ideas. This style of intellectual discourse became known as the "Copenhagen spirit," and inspired great breakthroughs in quantum theory. As Klein remembered those days, "In these surroundings it was natural to dream about the deeper background of these strange quantum rules with their whole quantum numbers."[5]

One time, Klein asked Bohr about ways of describing the motion of free particles through space-time. Bohr casually remarked that standard four-dimensional mechanics would likely prove inadequate for such a description and speculated that higher dimensions might do the trick. Klein took Bohr's remark at face value and began to think of ways of including extra dimensions in physics. Later he realized that he might have read too much into Bohr's statement. "I think his view was more the negative one that one could not have a theory in four dimensions," Klein observed in hindsight.[6]

The early 1920s proved eventful for both physicists. In 1921, Klein finally received his Ph.D. from Stockholm. Kramers was present at his thesis defense, so the discussion proved quite lively. Meanwhile, Bohr opened his own Institute for Theoretical Physics, a complex just beyond central Copenhagen that almost immediately became a mecca for modern science. The center, funded in part by the Carlsbergs of beer-brewing fame, was later renamed the Niels Bohr Institute. He was also honored with the first of what turned out to be many celebrations of his work, a series of lectures in Göttingen later called the Bohr Festival.

Klein accompanied his Danish mentor to the festival, where he was first introduced to Wolfgang Pauli and Paul Ehrenfest. Though fellow natives of Vienna, Pauli and Ehrenfest were also meeting each other for the first time. Pauli was glowing in recognition for a survey he had recently written on relativity, cosmology, and unified field theory, completed at the tender age of twenty. The treatise was so comprehensive that it was considered the definitive treatment of the field for decades. Even Einstein recognized Pauli as a wunderkind, calling his work profound, mature, and grandly conceived.[7] Ehrenfest, on the other hand, had just finished an original but controversial article on statistical mechanics, cowritten with his wife. Both articles appeared in the same encyclopedia.

Both Pauli and Ehrenfest shared the characteristic of being unusually blunt, as Klein was soon to find out. As Klein reported, "On that occasion Ehrenfest stood a little away from Pauli, looked at him mockingly and said: 'Herr Pauli, I like your article better than I

Wolfgang Pauli and Paul Ehrenfest. Two of the foremost twentieth-century Austrian physicists, each was outspoken in his views on higher dimensions.

like you! To which Pauli very calmly replied: 'That is funny, with me it is just the opposite!' "[8]

In 1922, one year after Einstein was similarly honored, Bohr received the Nobel Prize. At his Nobel lecture, Klein observed in apprehension as Bohr walked up to the podium without his notes; in his excitement, he completely forgot to bring them. Nevertheless, Klein was delighted to watch Bohr improvise a brilliant speech about the accomplishments and challenges on the road to a sound quantum theory.

Klein also had his own concerns at the time. In looking for an academic job, he hoped to stay in Sweden. He had a temporary offer at the University of Lund, but wanted something more stable. Meanwhile, Bohr discovered a visiting position at the University of Michigan in Ann Arbor and heartily recommended Klein for the post. After Michigan accepted his application, Klein agreed to the appointment, hoping to return to Sweden at some future date.

Just before setting sail for the States, Klein got married. The lucky bride was Gerda Agnete Koch, a student of Danish literature and the daughter of a physician. It amused them that while he was the son of a rabbi, she was related to bishops and priests. In biblical fashion, they would be blessed with six children.[9]

Ann Arbor Days

As Klein began his two-year instructorship at Michigan, starting in the fall of 1923, he needed to perfect his English. He was already fluent in Swedish, Danish, German, and French. Though he had learned English in school, he hadn't practiced it much before. Consequently, he amused his students with strange-sounding expressions such as "we must look apart from . . .". A supportive colleague, Walter Colby, would sit in his lectures and correct his mistakes. Students felt that this "spoiled their fun."[10] Klein's English improved considerably, until it became second nature to him. Eventually, his fluency would rival that of native speakers.

In the quiet town of Ann Arbor, isolated from the currents of European discussion, Klein found the opportunity to experiment with his own ideas about quantum theory. He began to examine connections between the quantum rules, which require integral multiples of fixed energies, and the properties of light. In optics, light waves can interfere (merge in additive fashion) with each other, generating predictable patterns of dark and bright bands. Under certain circumstances, these bands are spaced at equal distances, like the zebra stripes at a street crossing. Klein wondered if similar wave interference patterns could be used to understand the regular spacing of electrons in an atom.

Klein never published these ruminations—the first in his list of near-discoveries. An influential paper by the French physicist Louis de Broglie appeared at that time with similar views, but much more developed. In his Nobel Prize–winning research, de Broglie insightfully demonstrated how electrons (or any other subatomic bodies) have dual properties, resembling particles in some situations and waves in others. Relating electrons' momenta to their wavelengths and their energies to their frequencies (rates of vibration), he successfully predicted the patterns they would form if they interfered with one another. Furthermore, he found a clever way of explaining the positions of Bohr's electronic energy levels, by picturing the electrons as standing waves.

Standing waves occur whenever oscillations are confined to a finite space—for instance, air vibrating within a drum or pipe. Musicians rely on these to produce the fundamental vibrations and resonant tones that splendidly combine into lush sounds. A good example of this is the plucking of a guitar string. After the string is plucked, vibrations bounce off one end and the other. In short order, however, the original and reflected waves add up to a resultant vibration that oscillates up and down in place, rather than back and forth. These standing waves can have one peak, two peaks, or, in general, any whole number amount of peaks. One cannot imagine a string vibrating with one and three-quarter peaks, for instance. The natural modes of vibration constitute integer multiples of a fundamental frequency.

De Broglie proposed that an electron, circling the nucleus of an atom, acts as a vibrating standing wave. Like a plucked string, it can assume only integer multiples of a fundamental frequency, corresponding to multiples of a ground state energy. This explains why an electron can never be found in between the fixed quantum states. The integer values correspond to Bohr's quantum numbers.

Klein would later read de Broglie's paper with great interest, but also with a degree of skepticism. Though he agreed with the basic idea of treating electrons as waves, he felt that de Broglie failed to explain the underlying mechanics of how free electrons could move through space like particles. Somehow, Klein believed, collections of waves would have to add up just right to produce particle-like wave fronts. Then the particles would appear to ride through space like surfers perched on crests.

In order to construct just the right model of such a rolling wave front, Klein believed that three spatial dimensions plus time would not be enough. Therefore, recalling Bohr's advice, he decided to investigate five-dimensional models of particle motion. He hoped that by projecting five-dimensional motion onto four-dimensional space-time, the quantum properties would appear like variegated images from a prism.

As it turned out, there was indeed a way of understanding both the wavelike and particle-like behavior of the electron within the context of ordinary space and time, namely by means of the Schrödinger equation. But by the time it was developed, Klein had already spent close to two years probing the properties of five-dimensional theories.

The Closing Circle

In 1924, Klein's five-dimensional research began in earnest when he was teaching a course on electromagnetism. One of the problems he considered in class was the motion of an electron under the combined influence of gravitational and electromagnetic fields. At the time, he was by no means an expert on gravity, drawing much of his knowledge from Pauli's treatise on the subject.

Klein decided to attack the problem by means of a method known as the Hamilton-Jacobi equation. This involves defining the potentials (functions from which the field derives) for both the gravitational and electromagnetic fields. One then uses these potentials, along with the kinetic energy (energy of motion) expressions, to find the momentum (quantity related to velocity) variables associated with the position coordinates. In essence, it's applying a set of functions and an equation to tell you how fast something is moving if you know where it is—like using a weather model to find the wind velocity for Kansas under certain conditions.

After writing down the gravitational and electromagnetic potentials, Klein noticed interesting parallels between the two. Although the gravitational potential derives from geometric terms, namely Einstein's metric components, and the electromagnetic potential stems from a different basis, Maxwell's theory, Klein found that they played similar roles in the equation. By identifying the electromagnetic potential as the metric components for a fifth dimension and by designating the fourth component of the momentum to be a multiple of the electric charge, Klein realized that he could merge these terms into a uniform expression. It is like the climactic scene of the film *Vertigo*, in which the protagonist (played by Jimmy Stewart) establishes the true identity of a disguised woman by insisting that she restore herself to her original appearance. Similarly, Klein unmasked the electromagnetic terms to reveal their true identities as metric components. As he recounted this discovery: "Thereby the similarity struck me between the ways the electromagnetic potentials and the Einstein gravitational potentials enter into this equation, the electric charge in appropriate units—appearing as the analogue to the fourth momentum component, the whole looking like a wave front equation in a space of four dimensions. This led me to a whirlpool of speculation, from which I did not detach myself for several years."[11]

Klein then assumed, in a manner similar to de Broglie, that elementary particles behave like standing waves. He demonstrated that in such a case, a particle's momentum is inversely proportional to its wavelength (the bigger the momentum, the smaller the wavelength). The proportionality factor relating the two is a miniscule quantity known as Planck's constant. Because Klein identified the fifth dimension's momentum with the electric charge, that means that the electric charge is related, in turn, to the fifth dimension's wavelength.

Like de Broglie, Klein imagined these standing waves arranged into a circle. Only an integer number of peaks can fit into the circle's circumference. Because wavelengths constitute fixed fractions of the circumference and charge is related to wavelength, that implies that charge can come in only exact multiples of a particular unit. In the language of modern physics, that means that charge is quantized.

By Klein's day, physicists knew through oil drop experiments that the smallest free charge is that of an electron. Other charges are multiples of this fundamental quantity. Therefore, Klein's result elegantly reproduced a previously unexplained experimental fact.

Plugged into Klein's relationship between charge and wavelength, this smallest unit of charge yields the radius of the fifth dimension's circle. Klein determined this radius to be less than 10^{-30} inches in size—far, far smaller than even the minute proportions of an atomic nucleus. A trillion, trillion of these radii would only be the size of bacteria.

Klein found the minuscule scale of the fifth dimension to be most encouraging. He was pleased that his theory predicted that the fifth dimension could never possibly be observed. It provided an elegant explanation of why space-time appears to be only four-dimensional.

Envisioning a universe with a tightly curled-up fifth dimension is akin to trying to read a thinly rolled up scroll. Imagine taking a page from a book and wrapping it up snuggly until only parts of its sentences are distinguishable. Then picture rolling it up tighter and tighter until its words and then its letters all blend together. Eventually, as the page becomes twisted as thin as spaghetti, its print would merge into a gray smudge, impossible to discern at all. As in Klein's theory, the information would still exist but its presence in the rolled up direction would be blurred beyond perception.

Shipwrecked

During the summer of 1925, Klein and his wife bade farewell to the clapboard houses of Ann Arbor and set sail for the copper spires of Copenhagen. There, he completed his calculations, discussing them at length with Bohr. Bohr pointed out connections with de Broglie's thesis that Klein viewed with curiosity.

As the summer drew to a close, Bohr invited Klein to work at the institute on a fellowship. Embracing Bohr's generous offer, Klein

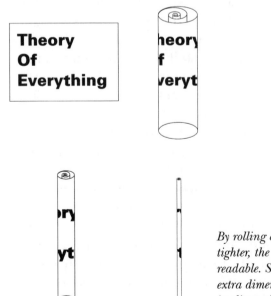

By rolling a piece of paper tighter and tighter, the words become no longer readable. Similarly, by curling up the extra dimension of a theory of everything, its direct effects become undetectable.

requested first to visit his mother in Stockholm, whom he hadn't seen for more than two years. Intending to stay in his native land only briefly, he eagerly anticipated returning to Denmark and presenting his completed five-dimensional model to his colleagues.

Unfortunately, poor health forced Klein to remain in Sweden much longer than expected. He was plagued by one ailment after another. First, he came down with the flu. Then, he became gravely ill with infectious hepatitis. Jaundice and fever confined him to bed for months. It wasn't until March 1926 that he was well enough to return to Copenhagen.

Upon his return to the Bohr Institute, Klein glumly realized that he had been sitting on the sidelines during a pivotal time in science. While he was ill, Heisenberg's matrix mechanics and Schrödinger's wave equation had appeared in succession, each describing the quantum behavior of subatomic particles and predicting the electronic structure of the atom. Klein had developed a version of the Schrödinger equation, but had not been able to publish it because of his infirmities. This was one of the greatest disappointments of his life.

To make matters worse, a few weeks after arriving back in Copenhagen, Klein found out about the similarities between Kaluza's theory and his own. Pauli, who was visiting the institute at that time, looked over his work, then gave him the bad news. Klein vaguely knew that Kaluza and Weyl had each proposed unified field theories, but he was unaware of the details. He had no idea that Kaluza's notion, like his own, involved extending general relativity by an extra dimension to accommodate electromagnetism. After Pauli pointed out the commonalities, Klein finally looked up Kaluza's paper. Indeed, though it was based on different methods and suppositions, it was uncannily similar in concept. Klein wondered if he had been wasting his time plowing through already tilled soil.

To ward off despair, Klein remembered some advice he had given Kramers during a particularly bleak moment in Kramers's life. Science, he had pointed out, should be treated like the play of children. If something goes wrong, one should just move onto a different game.

After cheering himself up, Klein resolved to publish his five-dimensional work anyway, trying, as he put it, "to rescue what [he] could from the shipwreck."[12] He decided to write two papers on the subject, emphasizing his own innovations. Even though he felt that his own theory had marked differences, he gave Kaluza full credit as the originator of the notion. Later, he came to regret this decision because it implied that his own work was derivative. As Klein revealed in an interview, "Kaluza . . . only derived the field equations in the first approximation. I had worked them out, in the summer before, rigorously. So I was not very impressed by Kaluza's paper, but thought that since he was much before me I should quote it. Nobody could see that I had it independently."[13]

Assigning the correct attribution to ideas became a sore point for Klein because de Broglie once wrote a paper designating Kaluza and Kramers as the cofounders of five-dimensional theory. Although Kramers had little to do with the notion, de Broglie misconstrued Klein's friendly references to Kaluza and Kramers in his first article to be a statement doling out full credit. Klein recoiled at the thought that two years of his original research would be ignored.

Fortunately for him, that was not to happen. Klein's papers, with their clear links to atomic theory, attracted greater notice in the physics community than Kaluza's work (and certainly far more than Nordström's obscure theory). Until then, only Einstein and his associates had expressed interest in Kaluza's five-dimensional idea. Klein's use of a higher dimension to interpret quantum effects appeared to be a reasonable alternative to other explanations advanced at the time. Moreover, his calculation of the miniscule size of the extra dimension—well in line with experimental limits—seemed a solid argument for his thesis.

Two St. Petersburg physicists, Vladimir Fock and Heinrich Mandel, independently developed ideas similar to Klein's around the time his first paper appeared. While their articles were in press, they found out about Klein's work. Wasting no time in assigning him proper credit, they became leading Russian advocates of five-dimensional theory. Fock's contributions were influential enough that Einstein sometimes called the approach "Kaluza-Klein-Fock."

Life in "Apartment 5-D"

One of the most enthusiastic supporters of Klein's ideas was Paul Ehrenfest. Deeply interested in the question of why space-time appears four-dimensional, Ehrenfest found Klein's explanation very

appealing. Moreover, Ehrenfest greatly respected Bohr and viewed
Klein as his brilliant young protégé.

To find out more about Klein's ideas, Ehrenfest asked Lorentz to
invite him to Leiden for a summer visit. When Klein received the letter from Lorentz, he was absolutely thrilled. Impressed since his
teenage years with the clarity and insightfulness of Lorentz's writings, Klein greatly looked forward to meeting one of his heroes.

In June 1926, Klein moved down to Leiden, sharing an apartment in a rooming house with one of Ehrenfest's young students,
George Uhlenbeck. Born in Java to a German-Dutch family, Uhlenbeck was a rising star in quantum and statistical physics. One year
before Klein's visit, Uhlenbeck, along with fellow student Samuel
Goudsmit, discovered the idea of electron spin. This finding proved
of critical importance to atomic theory, exciting Bohr, Heisenberg,
and many others.

The motivation for Uhlenbeck and Goudsmit's discovery came
from a suggestion by Pauli that the behavior of an electron in an
atom could be fully described by four fundamental quantum numbers, which classified all of its possible energy states. The first number was Bohr's primary energy levels and the second had to do with
geometry. The third pertained to the behavior of electrons in a magnetic field—called the Zeeman effect—but no one knew what the
fourth was. Uhlenbeck and Goudsmit cleverly surmised that the
fourth quantum number related to an additional way an electron
can move, namely its possibility of rotating either clockwise or counterclockwise. They referred to this novel concept as the *spin*. Pauli
duly noted that electrons grouped naturally into pairs of opposite
spins, rather than larger clusters, a notion known as the Pauli exclusion principle.

Klein's theory did not account for spin, a fact that did not bode
well. Nevertheless, Uhlenbeck was deeply in awe of his new colleague
and flatmate, believing that he had pointed the way to solving one of
the greatest riddles of all time. He "felt a kind of ecstasy" about the
grandeur of Klein's vision.[14] Uhlenbeck recalled one of the first
times he heard Klein speak about his model:

> I still remember one time after these discussions with Klein in which
> he had told about his five-dimensional relativity and how out of that

the quantum conditions would come. You see, from the periodicity condition in the fifth dimension you got the quantum conditions. And I was so excited. I told [my friends], 'Very soon we'll have the world formalized. We will know everything! Everything will be known at that time.' Well, it was a beautiful exaggeration.[15]

Uhlenbeck and Klein became fast friends, talking about five-dimensional theory almost every day during Klein's month-long visit. Naturally, Ehrenfest was involved in these discussions as well. He welcomed the chance to resume his research on dimensionality. In particular, he was curious to know why some methods work only in certain numbers of dimensions.

Unlike Uhlenbeck, however, Ehrenfest did not think Klein's theory would prove the panacea for all the ailments of physics. Although he was excited and impressed by Klein's work, he maintained a skeptical attitude toward all new approaches. It was simply in his nature to question everything. Still, he wrote in a hopeful note to Einstein, "I believe that his ideas—still momentarily so sketchy and precarious—could possibly lead to something better beyond Schrödinger."[16]

Ehrenfest, Klein, and Uhlenbeck spent so much time discussing these issues that they decided to write a collaborative paper on the topic. It attempted to explain the Zeeman effect and other quantum behaviors. But when the article turned out to be mainly Klein's ideas, Ehrenfest elected not to publish it. He didn't think it was fair to have his name on a paper than was primarily written by another. Later, after Klein returned to Copenhagen, Ehrenfest wrote a joint article with Uhlenbeck describing how de Broglie's waves behave in five-dimensional space. Though it referred to Klein's theory, the article represented Ehrenfest and Uhlenbeck's own graphical attempt to examine the behavior of five-dimensional waves.

The Great Schism

After bidding a sad farewell to Leiden, Klein resumed his research in a city that had become the undisputed center of modern physical thought. Bohr's institute had established itself as the Vatican of quantum physics, evaluating novel theories of nature, smoothing out differences between opposing factions, establishing canonical doctrine, and issuing Holy Writ. For a generation of pilgrims to the shrine of the atom, all roads led to Copenhagen.

The raging dispute of the times was between Heisenberg's and Schrödinger's proposed solutions to the quantum riddle. Heisenberg's approach was far more mysterious. While a bright-eyed youth of only twenty-three years old, he abolished the idea of fixed electron orbits, replacing them with unobservable mathematical abstractions. Calling his theory matrix mechanics, he demonstrated how special transformations between states could explain atomic structure.

Schrödinger's method, on the other hand, was more solidly planted in traditional physical theory. His equation described how de Broglie's matter waves either moved freely through space or vibrated in place while trapped in the potential wells of atoms. Like Heisenberg's model, the Schrödinger equation proved an extremely powerful tool, accurately predicting observed particle behavior and light spectra.

In June 1926, while Klein was in Leiden, physicist Max Born reinterpreted the Schrödinger equation in a radical new manner. He proposed that instead of describing matter waves, it delineated probability waves. The full properties of electrons and other particles, he stated, could only be known in a probabilistic manner, never exactly.

Schrödinger fumed at this transformation of his work. Wave mechanics, he felt, should be exact and deterministic, not fuzzy and probabilistic. "If you have to have these damn quantum jumps then I wish I'd never started working on atomic theory,"[17] he desperately told Bohr. In what would become a guiding principle, Einstein sympathized with Schrödinger's position. "I am convinced that the Old One does not play dice,"[18] he famously wrote to Born. Heisenberg, on the other hand, felt that Schrödinger's work was unsuitable in any form and rallied for a return to matrix mechanics. Nevertheless, all of these theorists were powerless to stop the inevitable incorporation of Schrödinger's work into a nondeterministic theory of atomic behavior; it was simply too successful a model.

In winter 1927, Bohr went up to the mountaintop in search of some inspiration. For him, skiing was a kind of religious experience, during which he could focus his thoughts in the crisp, cold air. He was struggling with how to reconcile the views of Schrödinger, Heisenberg, and others, melding them into a cohesive theory that everyone would accept. Finally, as the wind rushed past and the valley drew closer, he found the Zen answer he was seeking: the principle of complementarity. Both the wave and particle pictures were correct, he realized, but whenever an experiment brought one of these views into focus, the other faded into static.

He dashed back to Copenhagen, where Heisenberg was working as a researcher. Bohr excitedly told Heisenberg what he had discovered. Heisenberg, in turn, described to Bohr his own thoughts on the subject. While Bohr was away, Heisenberg had developed a notion, called the uncertainty principle, detailing how physical properties depend on observation. He proposed that many measurable features of particles, such as position and momentum, have paired relationships in which the more scientists know about one, the less they know about the other. Thus, complete knowledge of a particle's properties is impossible.

Klein observed these discussions with great interest. Although he was still hopeful about explaining quantum dynamics by the use of his five-dimensional theory, he had come to recognize the value of other approaches. It was hard to argue with experimental success, and both Heisenberg's matrix mechanics and the Schrödinger equation were adept at reproducing atomic spectral lines. Moreover, as an avid disciple of Bohr, he could not help but be impressed by the Danish physicist's sweeping vision of nature.

Witnessing Bohr and Heisenberg's debate about the merits of complimentarity versus uncertainty, Klein decided to act as a neutral broker. Through his diplomatic efforts, he helped convince them that both of their ideas were equally correct. After a number of long walks through the park to mull this over, Bohr and Heisenberg reached a pact. They agreed that the complementarity and uncertainty principles are two equivalent ways of understanding the same aspect of nature. Each addresses the symbiotic relationship between how an observer takes his readings and what he ends up observing.

Soon, persuaded by Bohr, Heisenberg came to accept that Schrödinger's formulation and his own were equally valid. Through much cajoling, Bohr pressed Schrödinger into agreement as well, at least for the time being. Finally, in a grand theological reconciliation, all the schisms that threatened to tear the quantum church apart seemed bridged.

Jocular Physics

The Niels Bohr Institute during the advent of quantum mechanics was not all serious business. There was ample time for levity and Klein certainly enjoyed his share. He loved composing humorous pieces to send to his colleagues, sometimes illustrated with cartoons.

In one satirical letter to Ehrenfest, Klein refers to him as "his majesty, the wisest, most powerful emperor in Leiden." Klein humbly calls himself Ehrenfest's "servant and consul to Bohrtown." He then proceeds to describe how "spring's buds and blooms tremble and bow before the strict majesty of the emperor, if he so orders." The letter is inscribed with Klein's hand-drawn "state seals" for Leiden and "Bohrtown."[19]

Ehrenfest responded with a letter in Latin addressed to "Consul Oscari Parvo in Danish Hafnia." He signs it "Paulus Honestus."[20] The

The Niels Bohr Institute in Copenhagen, where quantum theory was born.
Oskar Klein worked there for many years.

rather esoteric joke is that "Parvo" means "small" in Latin, the equivalent of "Klein" in German. Similarly, "Honestus" in Latin can be translated into "Ehrenhaft," or honorable. Finally, "Hafnia" is the Latin name for Copenhagen.

Not all of the humor was so obscure. Some of it was contemporary satire, with even a dose of politics mixed in with the physics. When Fascist dictator Benito Mussolini defied the League of Nations by resolving to annex Ethiopia, arrogantly proclaiming, "Italy will pursue her aims with, without or against Geneva!" Klein wrote a piece suggesting that world leaders should learn quantum physics. He advised them to make use of the principle of complementarity—with its union of opposites—by which "Bohr has been able to create almost complete harmony in the atomic world (including Pauli)."[21]

Klein submitted his letter, entitled "On political quantization," to a volume of humor dedicated to Bohr in honor of his fiftieth birthday. This *Journal of Jocular Physics* appeared only twice more—to commemorate Bohr's sixtieth and seventieth birthdays as well. Because of the risk of offending readers with its political message, Klein's piece was never published, however, and has remained in the Niels Bohr Archive ever since.

The most famous example of the institute's humor was a parody of *Faust,* produced and performed by its members in 1932. In it, Pauli was depicted as the diabolical Mephistopheles, Ehrenfest as the cynical Faust, Eddington as an archangel, and Bohr as the Lord himself. American physicists Richard Tolman and Robert Oppenheimer were seen downing drinks at "Mrs. Ann Arbor's Speak Easy." Einstein was portrayed as a king infested by fleas (unified field theories). There was even a ghostly "Quantum Walpurgis Night," satirizing Paul Dirac's theories. A script of the play, illustrated with Russian physicist George Gamow's clever caricatures, was presented to Bohr and the institute as a gift.

A Toast to the End of a Theory

During the late 1920s, Klein's five-dimensional theory receded into the background as he became increasingly involved with other projects. Along with physicist Pascual Jordan, a newcomer at the institute, he pursued the goal of a quantum theory describing fields. Klein had made up his mind "that one had to make a choice if one would be able to base quantum theory on the five-dimension approach, which would be a kind of causal theory in the fifth dimension, or if that were not possible there would have to be a similar treatment of the fields as the quantization in mechanics."[22]

Born in Hannover on October 18, 1902, Jordan owed his unusual first name to German-Spanish ancestry. He studied mathematics and physics in Göttingen with professors such as Hilbert and Born before coming to Copenhagen. In various collaborations with Born and Heisenberg—including the famous *Dreimännerarbeit* (Three-man work) of 1925 that established certain basic principles—he played a major role in the development of quantum mechanics. A shy, insecure man with a nervous stammer, he would come to feel that the physics community didn't award him proper credit for his contributions.

Jordan shared with Klein an interest in rendering the theory of quantum waves more rigorous, either through the five-dimensional concept or some other method. He was one of the few physicists at the institute who appreciated what Klein was trying to accomplish. His enthusiasm for Klein's model would lead him, much later in his career, to craft his own theory of unification. But at that point, neither Jordan nor Klein saw a way of pushing the five-dimensional approach any further. Therefore, as an alternative, they worked diligently together to find a way to quantize fields as well as particles. Their research led to an important procedure known as second quantization.

The final straw for Klein's five-dimensional aspirations in the 1920s (he would revisit the topic a decade later) came with a parcel from Dirac to Bohr. In it, Dirac enclosed the first draft of a paper detailing a relativistic quantum theory of electrons. Bohr was astonished by the brilliance of his work, which not only combined the two leading theories of the early twentieth century—relativity and quantum mechanics—in describing the electron, but also encompassed spin as well. In adding spin to the mix, he generalized a procedure developed by Klein, Walter Gordon, and others, which henceforth became known as the Klein-Gordon equation. (The Klein-Gordon equation is essentially a four-dimensional version of Klein's five-dimensional theory.)

In early 1928, hoping to find out more about Dirac's work, Bohr sent Klein to Cambridge, where the English physicist was a fellow of St. John's College. After conversations with Dirac about his succinct new equation, Klein came away a true believer. The Dirac equation, not the fifth dimension, seemed to him the road to a unified future. Klein returned to Copenhagen, reporting the exciting news to his mentor.

Then, in Easter of that year, Pauli arrived at the institute. Along with Heisenberg, he had also been trying to devise a quantum, relativistic description of the electron that included the spin, but had been beaten in the end by Dirac. There had even been a bet between Heisenberg and Dirac as to who would get to the finish line first. But it took no measuring tape for them to recognize Dirac's stunning victory.

During dinner one evening, Klein and Pauli decided to take out a bottle of wine. It was a new era for physics and a time for them to rethink their own research paths. They raised their glasses and made a toast: Here's to the death of the fifth dimension![23]

Throughout the year, Klein kept his word and abstained from any five-dimensional activities. Instead, he worked with Japanese physicist Yoshio Nishina on applying the Dirac equation to understanding a phenomenon called the Compton effect. Bohr and other institute members were very happy that their colleague had kicked the habit and was now a full believer in the standard interpretation of quantum mechanics. Heisenberg, who had been completely against the fifth dimension, was particularly glad that Klein "was no longer a heretic."[24]

On the other hand, at least one physicist in Russia was bitterly disappointed with Klein's new direction. In December 1928, Heinrich Mandel sent him a tearful note: "Mr. Gamow wrote to me recently that you have now completely renounced the idea of five-dimensional field physics, and have thereby found that it has all been a misunderstanding. It seems that all you have published about it so far was nearly always in greatest agreement with my own opinion. Therefore I'm sorry that I cannot concur with you anymore."[25]

In January 1931, Klein was appointed to a professorship in Stockholm, a position that he held until he retired in 1962. He relished the opportunity to return to his native city, where he and his wife could raise their six children in a peaceful setting—away, as it turned out, from the horrors that plagued other parts of Europe during the 1930s and 1940s.

Shortly before he left Copenhagen, Pauli, though younger than Klein, gave him this rather blunt "paternal" advice: "I hope you will now fulfill the words 'Go and teach the people.' Your great pedagogical ability was always one of your strongest suits. . . . I am not of the opinion that finding new laws of nature and indicating new directions is one of your great strengths, although you have always developed a certain ambition in this direction."[26]

The same winter that the Kleins headed north, the Einsteins began a westward voyage. Several prominent Caltech professors, including Robert Millikan and Richard Tolman, invited Einstein to Pasadena for a two-month research visit. It would be a chance for him to visit Mount Wilson Observatory, where the expansion of the universe had been discovered in the 1920s by Edwin Hubble. This was one of general relativity's greatest predictive successes.

Einstein didn't mind the chance to leave Europe for a while. Most of his fellow European physicists had gone "quantum crazy," as he saw it, believing in the magic of uncertainty. A devout determinist influenced by the philosophy of Spinoza, Einstein refused to sanction any role for chance in physical theories. When even one-time champions of causality such as Klein had been won over to quantum reasoning, Einstein became increasingly isolated from his peers.

While much of the physics community began to ignore his still prolific output, Einstein's admiration by the public seemed to reach no limit. He had become a living legend—the first international scientific superstar. Consequently, his writings were more likely to be featured in the popular press than to be seriously cited by his colleagues. Virtually any theory he developed or statement he made would attract the interest of newspaper reporters. He fully expected they would hound him during his trip.

Although this would be a research journey, he and Elsa looked forward to the pleasant climate of southern California and hoped to do some sightseeing. It was a land they had seen only in the movies, and soon they would be part of it as well.

CHAPTER 7

Einstein's Dilemma

Quantum theory is fully Schrödingerized and from this has come much practical success. But, nevertheless, this cannot be the description of the real procedure. It is a mystery. . . . It appears that the unification of gravitation and Maxwell's theory will be achieved in a completely satisfactory way by the five-dimensional theories of Kaluza, Klein and Fock.

—ALBERT EINSTEIN, in a letter to Hendrik Lorentz, 1927

The Tramp and the Professor

It was the time of the Great Depression, yet Charlie Chaplin's elegant Beverly Hills estate was always brimming with guests. Extraordinarily famous for his screen role as the Little Tramp, a vagabond with a derby hat, ragged suit, and cane, Chaplin was quite rich in real life and could afford to host lavish parties. His magnificent cinematic achievements—serving as an actor, writer, director, and even composer for numerous hysterical and heart-wrenching films—brought him exquisite fortune. Through the devastating stock market crash of 1929, Chaplin managed to hold on to his wealth by cleverly placing his money in stable accounts. Now, more than a year later, he continued to take delight in spending it on entertaining his vast circle of friends. Publishing giant William Randolph Hearst, screen idol Douglas Fairbanks, popular actress Mary Pickford, and numerous other celebrities often came to call, enjoying amenities in Chaplin's house such as "a Japanese theater, with genuine Japanese dances being performed by genuine Japanese girls."[1]

Chaplin was hardly superficial, however. In hosting visitors, he derived his greatest pleasure from intellectual conversation. The more learned his guest, the prouder he was. Therefore, when offered the chance to host Albert Einstein, a man attached to the word *genius* like Capone was to crime, Chaplin leapt to the occasion.

The Einsteins arrived by ship at the port of San Diego on December 30, 1930. They were treated to a hero's welcome—flower floats, beautiful "mermaids," and hordes of cheering onlookers. Seeing this spectacle on newsreels, their friends back home in Berlin thought that America had gone crazy.

The Einsteins were soon set up in a cozy gingerbread cottage in Pasadena, near the Caltech campus, where the physicist would be consulting with fellow scientists. "Here in Pasadena it is like paradise," he wrote to his friends. "Always sunshine and clean air, gardens with palms and pepper trees and friendly people who smile at one and ask for autographs."[2]

During the first week of the visit, Carl Laemmle, the German immigrant founder and head of Universal Studios, invited the professor to visit Hollywood. Einstein asked his fellow countryman for a special favor. The most famous scientist in the world wanted to meet the best known comic film star. Laemmle quickly arranged for Chaplin to join the Einsteins for lunch at the studios. Chaplin found Albert "jovial and friendly" and Elsa "a square-framed woman with abundant vitality."[3] The meeting went far too quickly, for it was soon time for the Einsteins to go on a studio tour. Consequently, Elsa took Chaplin aside and asked him directly for a dinner invitation. Rather than being shocked by her forwardness, Chaplin was delighted by the opportunity to get to know the Einsteins even better, and immediately began preparing for the occasion. A relaxed dinner at his house, Chaplin anticipated, would be the perfect opportunity for the renowned thinker and him to share extraordinary tales with each other.

Before swapping stories about the blessings of creativity and the hazards of fame, Chaplin and Einstein needed to face some communication issues. Though Chaplin could fake a credible German accent, throwing in a few familiar words and phrases for comic effect, he couldn't really speak the language. When engaged in serious discussion, he spoke in a quiet, measured English—a mood and a voice the public rarely experienced.

Einstein, on the other hand, on only his second trip to the United States, could barely speak English at the time. He delivered his lectures exclusively in German. Fortunately, he could count on Elsa, who spoke English very well, to translate if needed for social situations.

This helped alleviate one language barrier, but there was a second they needed to address. Chaplin couldn't speak physics—a language in which his dinner guest was extraordinary. To head off any communication problems in that area, he brought along his own "translator," Dr. Cecil Reynolds, who fancied himself knowledgeable in the field. But Dr. Reynolds wasn't a physicist. Rather, he was Chaplin's personal physician, a brain surgeon who moonlighted as an actor and hobnobbed with Hollywood types. Brain surgery, he once told Chaplin, "is merely knowing where the nerve fibers lie, but acting is a psychic experience that expands the soul."[4]

Later in 1931, as a medical consultant, Reynolds would be technical advisor to the film *Frankenstein,* helping to intensify the realism of its depiction. In 1936, he would play the bit part of the jailhouse minister in Chaplin's *Modern Times.* Among his various talents, Chaplin felt that Reynolds knew a "smattering of physics." That was more than Chaplin's other friends, and served as his ticket to dinner with the founder of relativity.

During the course of the evening, Reynolds had the chance to show off his knowledge. He brought up a recent book he had read, *An Experiment with Time* by J. W. Dunne. The 1927 treatise used concepts in relativity to justify the possibility of communication between past, present, and future—including prophetic dreams and other claimed psychic experiences. Had Einstein read it? No. Einstein never read speculative works aimed at the general public—certainly nothing connecting science with spiritualism.

Reynolds proceeded to summarize parts of the book and ask Einstein his opinions about Dunne's ideas. "He has an interesting theory about dimensions, a sort of a . . ." Reynolds described, suddenly realizing he was in over his head. "A sort of an extension of a dimension."[5]

Einstein was amused. Constantly barraged by people with all different perspectives who had questions about all things under the sun (and beyond), he maintained a ready sense of humor. "An extension of a dimension . . . what's that?" he whispered mischievously to Chaplin.[6]

Reynolds quickly switched the topic and asked Einstein whether he believed in ghosts. No, Einstein responded, he had never seen one. If at least a dozen credible witnesses saw one at the same time, then maybe he'd believe. Until then, he would remain skeptical. What about levitation and other psychic phenomena? Could a trained mind raise tables high up in the air solely through the power of thought? Einstein shook his head. Without solid scientific proof, he could not lay credence to such things.

Reynolds's rapid segue from higher dimensions to the occult was symptomatic of the long-standing public association between the two topics, dating back to Zöllner, Slade, and the Society for Psychical Research. No wonder Einstein often felt the need to clarify this distinction whenever he mentioned higher-dimensional theories.

Finally, changing the subject back to physics, Chaplin asked Einstein if his theory of relativity contradicted the work of Isaac Newton. Einstein replied that, on the contrary, it augmented Newton's ideas. In saying this, Einstein emphasized, as he did on many other occasions, that he believed strongly in an objective, mechanistic universe. On the two hundredth anniversary of Newton's death, he wrote, "May the spirit of Newton's method give us the power to restore unison between physical reality and the profoundest characteristic of Newton's teaching—strict causality."[7]

Nothing, he felt, should be accepted as fact if it couldn't be verified again and again by independent observers. Claims of supernatural powers, for instance, failed that test. The Copenhagen interpretation of quantum mechanics (as the views of Bohr, Heisenberg, and their colleagues became known) asserting that atomic measurements vary depending on how observers take them similarly did not live up to Einstein's standards. There must be some deeper way, he felt, of modeling atomic behavior such that experiments would always yield the same unequivocal results. This strict adherence to the principle of objective experimentation and the rejection of the uncertainty principle were tenets from which Einstein would never waver, no matter how isolated he became from the mainstream physics community.

As an iconoclast himself, Chaplin greatly respected Einstein's independent spirit. In an age of talkies, Chaplin was the only director with the patience and courage to continue to produce silent films. It took boundless energy to round up actors who were even

Albert Einstein and Charles Chaplin. Einstein was Chaplin's guest at the January 1931 premier of City Lights. *Just as Chaplin sought to understand Einstein's groundbreaking theories, Einstein was interested in Chaplin's poignant depiction of the human condition. Both were fiercely independent, resisting trends as they shaped their creations.*

willing to do pantomime and train them in the precise gestures of sight comedy. Sometimes it would take days just to perfect a minute of footage. Other filmmakers couldn't understand why he went through all the trouble. But, like Einstein, he wanted to continue a sacred tradition, no matter what his colleagues said.

Einstein thoroughly enjoyed his visit to the Chaplin estate. He reported that Chaplin was "an enchanting person, just as in his film parts."[8] They were to meet again on a number of occasions in California as well as Berlin. Despite Reynolds's ramblings about the supernatural, the dinner party, laying the groundwork for a great friendship, was a success after all.

Battle of the Titans

It is ironic that Einstein felt so alienated by quantum theory, for in its early days he was one of its great pioneers. His Nobel Prize officially commemorated his discovery of the photoelectric effect, which demonstrated that light came in small packets, or quanta. This was one of the major advances that would lead to the concept of wave-particle duality. Later, he was the codeveloper of quantum statistics and the initiator of theories regarding wave mechanics. Moreover, he had the highest regard for many of the theory's founders, including Bohr.

Bohr and Einstein first met in 1920, when Bohr delivered a lecture in Berlin on atomic theory. Ehrenfest, close friend to both, recommended each to the other in his typical matchmaking style. After Bohr returned to Denmark, Einstein wrote to him saying, "Rarely in my life has a man given me such joy by his mere presence as you have. I understand now why Ehrenfest loves you so much."[9] Bohr returned the compliment, writing, "It was for me one of the greatest experiences I have ever had to meet you and to speak with you."[10]

When de Broglie's work appeared, Einstein strongly supported it. He saw it as the seed of a new approach to atomic physics. He was similarly impressed with Schrödinger's equation in its original incarnation as a causal description of matter waves. In May 1926, he wrote to his friend Besso, "Schrödinger has produced a couple of marvelous works about quantum rules. It smells of deep truth."[11] As it turned out, that was one of Einstein's last wholehearted expressions of approval for quantum mechanics.

One month later, Max Born reinterpreted Schrödinger's material waves as probability waves and Einstein drew his line in the sand. Einstein firmly believed that all events in the universe flowed from their antecedents with the absolute predictability of night following day. Anything that seems random must have resulted from a causal sequence that we have not yet understood. Either the data is incomplete or the physics is incomplete. A being possessed with the full awareness of any point in the universe's history, including a complete set of information and perfect knowledge of the laws of nature, would be able to forecast all events in the future.

Einstein saw the history of creation as a unified network of occurrences. Like Ariadne's thread, it stretches unbroken from the beginning of time until the end. Therefore, from the space-time perspective, the past and future are just as real as the present. "To us believing physicists," he would later write, "the distinction between past, present and future has only the significance of a stubborn illusion."[12] How, then, could there be any room for randomness? The good Lord, as he told Born at that time and would soon emphasize to many others, does not roll dice to determine the future—the future is already written.

In January 1927, Einstein wrote to Ehrenfest with his second thoughts about the Schrödinger equation: "My heart will not warm up to Schrödinger's work—it is not causal and is in general too primitive."[13] By then, Einstein had clearly dug in his heels. His friends knew where he stood and soon he would reveal his thoughts to the world.

During the course of that year, the northern lights were glowing with pride. In Copenhagen, none could be happier about the new quantum success story. With fatherly admiration, Bohr marveled at the achievements of the talented young researchers at the institute and was even more enthralled that they were beginning to speak with a unified voice. Thanks to Klein's help and plenty of fruitful discussions, he and Heisenberg were on the same page. The twin pillars of the uncertainty principle and complimentarity, Bohr believed, would provide the philosophical foundations for a new edifice of atomic theory.

In October 1927, Bohr proudly unveiled the Copenhagen interpretation of quantum mechanics during an international conference at the Solvay Institute in Brussels. (It was a lecture he had delivered once before, in Como, Italy, so it is generally known as the Como Lecture.) His audience was a who's who of modern physics, including Heisenberg, Schrödinger, de Broglie, Born, Ehrenfest, Lorentz, and—listening attentively to every word—Einstein.

Einstein responded immediately after Bohr finished speaking. The unexpected sharpness of his rebuttal suddenly changed a celebration into a debate. Complementarity was unacceptable, he argued, because it left particle behavior up to chance. Strict causal laws, not probabilistic procedures, must be found to explain all natural phenomena.

The audience became a ruckus of dissonant voices, arguing in a dozen languages about Einstein's words. Lorentz, presiding at the meeting, tried in vain to restore order. Finally, Ehrenfest strode up to the blackboard and jotted down a quote from the Bible: "The Lord did there confound the language of all the Earth." Everyone erupted with laughter at the reference to the Tower of Babel.[14]

From that point on, the Solvay conference became a boxing match between two veteran fighters. Einstein would come up with example after example of why Bohr's interpretation could not be complete, and Bohr would knock them down one by one. Ehrenfest acted as the referee, mediating the debates. As Heisenberg later described the action:

> The discussion would usually start at breakfast, with Einstein serving us up another imaginary experiment by which he thought he had definitely refuted the uncertainty principle. We would at once examine his fresh offering, and on the way to the conference hall, to which I generally accompanied Bohr and Einstein, we would clarify some of the points and discuss their relevance. Then in the course of the day we would have further discussions on the matter, and, as a rule, by suppertime we would have reached the point where Niels Bohr could prove to Einstein that even his latest experiment failed to shake the uncertainty principle. Einstein would look a bit worried, but by next morning he was ready with a new imaginary experiment more complicated than the last. . . . After the same game had been continued for a few days Einstein's friend Paul Ehrenfest said: "Einstein, I am ashamed of you; you are argu-

ing against the new quantum theory just as your opponents argue about relativity theory." But even this friendly admonition went unheard.[15]

Einstein arrived at the Solvay conference a respected leader of the physics community. He walked away from the meeting a loner, not a leader—still venerated for his earlier work but no longer valued for his new ideas. He would neither be the first nor the last to be left behind by "progress."

Three weeks before the Solvay conference, an event took place that was as monumental to the world of cinema as the meeting was to the world of physics. To great fanfare, *The Jazz Singer,* the world's first talking movie, premiered on the silver screen. While physicists in Brussels heralded the marvels of the new quantum order, filmgoers in New York queued to experience a different kind of miracle: hearing and seeing the recorded performances of movie stars.

Suddenly, the classic methods of filmmaking were completely outdated. Practically overnight, thousands of movie theaters were wired for sound. Once they witnessed talking and singing on screen, audiences wanted nothing less. As later parodied in the film *Singing in the Rain,* a whole community of silent performers found itself out of work. Only one director, Chaplin, had the clout and willpower to continue making silent films for the next decade, culminating in masterpieces such as *City Lights* and *Modern Times.* He believed strongly in his craft and didn't want to give it up, no matter what his colleagues advised him to do.

Einstein had the same single-minded independence as Chaplin. He was the only prominent physicist of his day to reject the quantum approach and continue to espouse classical determinism. He argued wholeheartedly that the universe is entirely predictable, no matter what the experiments seemed to indicate. He persisted in thinking, despite the opposition of most of the physics community, that quantum phenomena must form part of a grander, fully causal scheme. In doing so, his endeavors increasingly became a pantomime in an age of intense vocal discussion. Einstein's image persisted, but his words began to fall on empty ears.

To fight for his beliefs, Einstein drew upon what he saw as his most powerful arsenal: his ability to construct a unified field theory so comprehensive that it would include quantum effects as a natural consequence. To that aim, he devoted virtually the remainder of his scientific career, pushing himself to the limits of his health and energy. His quixotic mission to find the order inherent in all physical systems ended only when his fragile body could support it no more.

During his trip to California, Chaplin proudly took Einstein to the Los Angeles premiere of *City Lights*. Einstein was moved by the bittersweet saga of a poor but selfless tramp helping a blind flower girl. The ending of the film brought tears to his eyes. The silent screen character with the rumpled clothes and the mustache—so alone and so misunderstood—never gave up. Neither would Einstein.

Long Live the Fifth Dimension

In his isolation, Einstein persisted with boundless devotion. He saw two possible roads to the promised land of unification. One, paved by Weyl and Eddington, involved tinkering with the underlying geometry of general relativity. The other, laid out by Kaluza and Klein, required extending reality by an extra dimension. Although the latter path seemed more promising, it also seemed fraught with peril. How could he assume the existence of a fifth dimension that no one had ever observed? The cylinder condition, he felt, was so arbitrary. Progress in this field would require a more substantial explanation. Yet he couldn't give up. That would be ceding ground to Bohr. Thus, approaching the age of fifty, when many theorists contemplate retiring from active duty, Einstein was faced with an unusual dilemma: how to carry out what he saw as perhaps the most pivotal stage of his research career.

He had to make a decision. In January 1928, the answer momentarily seemed clear. Exuberantly, he wrote to Ehrenfest, "I think that Kaluza-Klein has correctly indicated the right way to proceed. Long live the fifth dimension."[16]

And so, with that affirmation, Einstein started step by step down the yellow brick road of theoretical pursuit—but not to a five-dimensional Oz just yet. Somewhere along the way to his goal, he got lost in the forest of second thoughts. Temporarily setting aside his affirmation to extend Kaluza's work, he attempted yet another varia-

tion of a non-Riemannian geometric approach called distant parallelism. Distant parallelism turned Weyl's theory on its head by keeping lengths invariant but changing the definition of parallel lines. In
this strange world, parallelograms no longer close; rather, they gape
like broken window frames.

Weyl, Eddington, and many other physicists found distant parallelism to be a disturbing rejection of relativity's major accomplishments. Pauli was especially critical, wondering why Einstein would
devise a theory that failed to predict correctly the bending of light
and the procession of Mercury's orbit. Was Einstein throwing all his
other accomplishments out the window for the sake of proving that
his research was still vital?

The press, on the other hand, loved the image of the graying,
shaggy-haired professor still hard at work trying to explain the universe. Because the publication of his new method nearly coincided
with his fiftieth birthday, Einstein received more publicity about it
than for any of his other unified approaches. However, this was naive
public attention, not the respect he sought from fellow scientists. In
January 1929, the *New York Times* announced the theory on its front
page, suggesting that it could be more important than relativity.[17]
Three weeks later, another *Times* article reported that all the publicity was driving Einstein crazy and that he was seeking seclusion from
the press.[18]

He subsequently bought a parcel of land in Caputh, a village
near Potsdam, and built a country house for Elsa and himself. He
went sailing in the lakes nearby, in a boat that was a birthday present.
There, feeling like a vagabond, he enjoyed his solitude and peace-
fully contemplated aspects of his unification proposal.

By that point in his career, Einstein had begun the practice of
having a research assistant perform the mathematical calculations
for his projects. He preferred to spend his time harvesting new ideas
rather than grinding them through the mill of equations. Otherwise,
the endless procession of mechanical details would have sapped his
creativity.

Einstein's "calculator" during that period was Austrian physicist
Walther Mayer. Already in his forties, Mayer had considerable experience at various universities, including Paris, Göttingen, the ETH,
and Vienna (where he received his Ph.D.) before coming to work for
Einstein in 1930. A rather unassuming, diminutive figure, he quietly

retreated into the background whenever he accompanied Einstein to events.

To provide additional help, Einstein had a superb secretary, Helen Dukas. Dukas had wonderful organizational skills—essential for keeping track of Einstein's steady stream of appointments whenever he was ready to receive them or shielding him from visitors whenever he wasn't. She would remain his secretary until his death and keep his papers in order for some time afterward.

Both Mayer and Dukas accompanied Einstein on his 1931 trip to California. Einstein would consult with Mayer in between his obligatory talks and meetings at Caltech. Although they began their collaboration by working on a project related to distant parallelism, Einstein was beginning to have sincere doubts about the validity of the theory. One of his hosts, physicist Richard Tolman, reported, "Einstein has been very, very kind to everyone. He has talked about his unified field theory twice and we are all impressed by the intelligence with which he goes at the problem even though he himself says it all may be a 'soap bubble'. "[19]

Concealing the Evidence

When, during the dinner with Chaplin, Reynolds asked Einstein about "extensions of dimensions," Einstein seemed puzzled and amused. Yet, ironically, that was precisely the sort of model he would work on next. It would be his first truly original foray into the world of Kaluza-Klein theory.

Einstein's very first paper on higher dimensional theories, written with Jakob Grommer in 1922, had reproduced Kaluza's model without the stress-energy tensor and had failed to find reasonable solutions. A second work, published in 1927, had echoed Klein's ideas without providing the quantum connection. By 1931, with the collapse of distant parallelism's house of cards, Einstein was ready to return to the subject, this time with more ambitious intent.

One of Einstein's great worries about higher-dimensional theories was introducing new physical properties that could never be observed. All consequences of any natural model, he felt, should be subject to experimental verification. As Einstein later wrote with colleague Leopold Infeld, "Physical theories try to form a picture of reality and to establish its connection with the wide world of sense impressions. Thus the only justification for our mental structures is whether and in what way our theories form such a link."[20]

Consequently, in constructing a new Kaluza-Klein model, Einstein did not want to propose anything "ghostly" that could never be detected. Because a physical extra dimension could not be tested directly, he and Mayer excluded it and went for something subtler. Like a judge determined to keep an offender behind bars, they deliberately designed a theory in which the extra dimension couldn't escape from the equations and be perceived as real. They made it a mathematical, not a physical, extension of the other dimensions.

As Einstein and Mayer described their approach in a paper on the subject, it was a formalism "which psychologically links up with Kaluza's well-known theory, while at the same time avoiding the extension of the physical continuum into one of five dimensions."[21] In this manner, the researchers hoped to circumvent the otherworldly aspects of dimensions beyond space and time while preserving their mathematical usefulness for a unified approach.

The technique they used involved keeping space-time four-dimensional, while assigning an abstract five-dimensional vector space to each point. One can think of this as a higher-dimensional bundle of arrows protruding from each of an endless array of targets. The arrows interact in the abstract space, with the real space-time feeling only the indirect repercussions. Because the direct action remains outside of our own domain, we can never sense the presence of extra dimensions.

As promising as this theory sounded, Einstein and Mayer soon realized its limitations. It replicated none of the features of quantum mechanics and would thereby be of little use as a rebuttal to Bohr. To make matters worse, it didn't accurately model the classical behavior of particles without recourse to unnatural assumptions. And it failed to produce matter from geometry, one of Einstein's major goals. Once again, without embarrassment or regret, he decided to move on to a new strategy.

Indeed Einstein went back and forth so many times on the issue of higher dimensions and other matters related to unified field theories that in late 1931 he was savaged by Pauli:

[Einstein's] never-failing gift of invention together with his stubborn energy in pursuit of a definite goal has recently bestowed upon us, on the average, one such theory per year—in which connection it is psychologically interesting that the current theory is usually depicted by its author for a period of time as the "definitive solution." So, in a variant of the well-known historical saying, one might exclaim at the appearance of a new attempt on this subject: "The old theory of Einstein is dead. Long live the new theory of Einstein!"[22]

Einstein thought long and hard about Pauli's comments. Pauli's barbs were sharp and venomous, but they were often right on target. Moreover, although he could be extremely critical, Pauli was one of the few quantum physicists who would actually read over Einstein's papers and make suggestions. Several months later, after trying one last time to get his latest theory to succeed, Einstein wrote to Pauli, "You were right after all, you rascal."[23]

A Captivating Offer

In the winter of 1932, Einstein visited Caltech once again as part of a new annual arrangement. There, he gave a number of talks focusing on unified theory and cosmology and consulted with Dutch astronomer Willem de Sitter about aspects of the expanding universe. He also spoke at length with scientific administrator Abraham Flexner, who was planning to establish a new think tank in New Jersey where scientists could work freely on their own projects without the distractions of university life. A large donation that he had obtained would pay the expenses. Einstein was fascinated by Flexner's proposal and agreed to resume their discussions later.

Upon returning to Germany, he stayed only two weeks before heading over to England to meet with Eddington and others. In Oxford, Flexner visited Einstein and provided him with further information about the new research center. The Institute for Advanced Study, as it would be known, would be located at Princeton but remain independent from the university. As Einstein's bright eyes glowed warmly at the thought of a refuge where he could work on his unification theories undisturbed, Flexner mentioned the possibility of a position.

In further meetings later that year in Caputh, the two of them ironed out the details. Einstein would spend five months of each year in Princeton, presumably replacing his Pasadena visits. Although he humbly asked for very little, Flexner guaranteed him a good salary, travel expenses, and even coverage of his taxes. Einstein enthusiastically agreed to the conditions. As Flexner boarded the bus out of town, he left with the unmistakable impression that everything was settled.[24]

But a few days later, delivered to Flexner in a thank-you note, came the sticking point. Before he would accept the appointment, Einstein stipulated that Mayer be awarded the position of full-time associate. "Now my own wish is that Dr. Mayer, my excellent coworker, will receive an appointment that is formally independent of my own," Einstein wrote. "Until now he has suffered very much from the fact that his abilities and achievements have not found their deserved recognition. He must be made to feel that he is being appointed because of his own achievements and not for my sake."[25]

When Flexner balked at such an arrangement, Einstein threatened to accept another offer. In a follow-up letter to Flexner, Einstein wrote of a position in Spain he was considering that would allow him to bring Mayer with him as another full professor. Flexner got the hint and reluctantly agreed to Einstein's demands.

Fleeing the Reich

Since World War I, the economic situation in Germany had become increasingly dire. Massive unemployment and devaluation of the currency drove the population to desperate measures. Exploiting these circumstances through religious, racial, and political scapegoating, the far right found the opportunity to seize increasing political power.

Einstein's sympathies were on the left and he was hopeful that the majority of the public would ultimately discount the Nazis' hateful propaganda. For a time, he and Elsa optimistically believed that they

Albert Einstein and Walther Mayer. Mayer served as Einstein's
"calculator" in attempts to develop a unified field theory.

could ride out the storm and remain in Caputh for at least part of each
year. He lent his name to an antifascist campaign, a union of Social
Democrats and Communists, that sought to prevent a Nazi victory.

Then one horrendous blow came after another. In July 1932, the
Nazis won a political plurality. Hitler was not yet in power, but the
gathering storm cast its shadow upon all who cherished liberty and
diversity. Soon, a reign of fire and a thunder of marching jackboots
would envelope much of Europe, laying waste to a vibrant intellectual culture. Virtually an entire generation of independent-minded
thinkers would have to flee from the burning house if they could or
else perish in the conflagration with millions of murdered victims.

The news of Hitler's accession to power came when Einstein was on his third trip to Pasadena, in the winter of 1933. Like his other sojourns, he was having productive discussions with Tolman and pleasant times with his friends. Chaplin hosted him several times. Once, Einstein surprised his host by bringing along three musicians and performing in a Mozart quartet. Chaplin delighted in seeing the professor play the violin so passionately. During another visit with Chaplin, a dinner party with celebrity guests, Marion Davies, the notorious mistress of William Randolph Hearst, brazenly ran her fingers through Einstein's famously unkempt mane, saying, "Why don't you get your hair cut?"[26]

The merriment ceased when Einstein discovered what had happened in Germany. As a Jew and a socialist known for his opposition to fascism and militarism, he knew that he could never return while Hitler was in power. Therefore, his trip back from California to Europe would need to avoid German soil at all costs.

Arriving by ship in Belgium, Einstein took steps to sever his ties with Germany. He resigned from the Prussian Academy of Sciences and renounced his German citizenship. One of the secretaries of the academy, a Nazi sympathizer, responded with a statement on its behalf condemning Einstein. Disgracefully, with the exception of Max von Laue, none of the other academy members disputed this resolution. Virtually overnight, the academy's darling had become its whipping boy.

Renting a cottage in the seaside town of Le Coq sur Mer, Einstein contemplated his options for the future. Fortunately, unlike many of his colleagues, he had a wealth of possibilities. He received offers from Oxford, Madrid, and many other universities. Caltech's physics department, under Millikan, was especially interested. Then, Flexner invited him to extend his planned stay at the Institute for Advanced Study for the whole year. Somewhat overwhelmed, Einstein pursued negotiations with several different places, waiting until the last possible moment to make his decision.

Although Caltech was tempting, Einstein ended up pursuing Flexner's generous proposal. As historian Gerald Holton relates, "The reason is that he knew that at Caltech he would be paraded by Millikan and others, as he had been on his three previous visits—partly to do fundraising—whereas at Princeton they would leave him alone. And he wanted to be left alone. So he chose Princeton and was very happy there—to be allowed to go in his sloppy sweater, not to have to dress up in evening clothes for yet another banquet for the Orange County millionaires."[27]

Requiem for Ehrenfest

Einstein headed for the United States by way of England. Along with Helen Dukas, Walther Mayer, and Elsa, he would sail out of Southhampton to make what turned out to be his final overseas voyage. He would never see Europe again.

While Einstein was in England, he was shocked to learn of the tragic death of one of his dearest friends. On September 25, 1933, Paul Ehrenfest, overwhelmed with depression, shot one of his sons, then took his own life as well. The extraordinary teacher and scholar had convinced himself that his own existence was worthless.

The reasons for Ehrenfest's suicide were rather complex. He had a lifelong sense of inferiority, believing that he could never match the achievements of his colleagues. In particular, he often felt that his appointment to the Lorentz chair at Leiden had been a great blunder. How could he ever fill the shoes of one of the most famous scientists in Europe?

He compared himself not only to Lorentz, but also to Einstein and Bohr. The blessing of having best friends who founded entire fields of thought proved a curse for someone prone to supreme feelings of inadequacy. Being surrounded by other Nobel Prize recipients didn't help, either. Furthermore, faced with rapid advances in quantum theory, he felt increasingly left behind.

Ehrenfest's personal life was in shambles. He had become estranged from Tatyana, finding affection in the warm embrace of an effusive young woman with artistic proclivities. Trading his wife's intellectual companionship for the nurturing he craved, he tried to relieve some of his insecurities. Nevertheless, as an honest and upright man, having an affair proved to be a great emotional strain. In the end, he hoped for reconciliation with his wife but couldn't see how to go about it.[28]

Then there was the matter of Vassik, Ehrenfest's youngest child, who had Down's syndrome. A number of years earlier, Ehrenfest had placed him in a well-regarded facility in Jena, Germany. But with Hitler's coming to power, he was rightly terrified by how the Nazis would treat a half-Jewish boy with a disability. Rescuing Vassik from Germany, Ehrenfest moved him to an institution in Amsterdam. This child care was very expensive, another source of worry.

Crushed by a huge financial burden, Ehrenfest greatly hoped that he, like Einstein, could find a new position in the United States. He contacted Tolman and other American friends, asking them for help obtaining any kind of job, be it in California or on the East Coast.[29] They put out many enquiries, but could not land him a suitable place.

As Ehrenfest's biographer Martin J. Klein relates, "During the last year or more of Ehrenfest's life he was struggling to find a way out of all these what seemed to be totally impossible situations. He looked very hard to find a position that would pay him very well. Finally he saw no way out. He thought he would make his position available and remove the burden to his family."[30]

With all hope lost, Ehrenfest wrote a farewell letter to Einstein, Bohr, Tolman, and some of his other friends, expressing how unbearable his life had become and apologizing for what he was about to do. The note was never delivered. He then met Vassik in the waiting room of his institution and killed him—apparently to relieve the rest of the family of the burden of support. Finally, Ehrenfest took his own life.

In a moving tribute, Einstein praised his late friend's impassioned teaching style and selfless encouragement of students. Perhaps projecting his own concerns, he speculated that the suicide was due to the problems Ehrenfest had fitting in at school and the difficulties of turning fifty. Neither he nor any of Ehrenfest's friends had grasped the full picture until it was too late.

For Einstein, the sorrow could not be more profound. First, his country was lost, now his own "brother" as well. As his ship sailed on to the New World, only the oceans of time distancing him from his losses could ultimately ease his pain.

CHAPTER 8

Truth under Exile

Theorizing at Princeton

Einstein was motivated not by logic in the narrow sense of the word, but by a sense of beauty. He was always looking for beauty in his work. Equally, he was moved by a profound religious sense fulfilled in finding wonderful laws, simple laws in the universe. . . . I asked him once about a theory and he said, "When I am evaluating a theory, I ask myself, if I were God, would I have made the universe in that way." If the theory did not have the sort of simple beauty that would be demanded of a God, then the theory was at best only provisional.

—BANESH HOFFMANN, *Working with Einstein*

Peter Bergmann once told me that there was always a first phase [of developing a unified field theory] when Einstein was very uncritical. Then, like a gardener, he would pick up the plant and look at its roots. After examining it closely, he would then have a more critical attitude. A few days later he would have a new theory. He would progress through these moods—uncritical, then self-critical, then uncritical—again and again.

—JOHN STACHEL, physicist and historian

Subtle, but Not Malicious

When Einstein arrived in the United States in October 1933 for what turned out to be his permanent relocation, his entrance was far more subdued than on his other visits. Abraham Flexner arranged for him to be picked up directly from the ship and whisked, in clandestine fashion, from the New York immigration processing center directly to Princeton. Flexner was nervous something would happen to his prize catch. The mayor of New York, who had hoped to greet Einstein with a speech and a parade, was left abandoned like a jilted bride.

Once Einstein arrived in Princeton, he could have the pleasure of warming himself by a personal fireplace that preceded him by a few years. The fireplace and indeed the whole new building, where the math departments of the institute and university shared space until the institute could build its own quarters, were meticulously planned out by Princeton's imaginative dean of Science, Oswald Veblen. Veblen envisioned the complex, known as Fine Hall, as a cozy community of scholars exchanging abstract ideas while sipping thought-stimulating cups of tea. Therefore the center prominently featured a tearoom, close to all of the offices. Years earlier at a seminar, Veblen had overheard Einstein casually remark, "God is subtle (in the sense of crafty or tricky) but not malicious." When designing the tearoom, he wrote to Einstein and obtained permission for that phrase (in its original German) to be carved above its fireplace. Little did he anticipate at the time that the scribe would soon follow the saying as a fixture of the same faculty lounge.

For good reason, Veblen was highly curious about Einstein's work on unified field theories, particularly his papers with Mayer on a variation of Kaluza's theory. The Einstein-Mayer articles greatly resembled work Veblen had published in 1930 along with his graduate student Banesh Hoffmann. Like the Einstein-Mayer model, the Veblen-Hoffmann work, known as *projective relativity,* involved a fifth dimension that extended into an internal, mathematical space rather than an external, physical realm. In this manner both models performed the magic trick of turning five dimensions into four, thereby avoiding the difficulty of assigning physical meaning to the extra dimension. In 1931, Veblen had pointed out the similarity in a polite letter to Einstein: "The New York Times this morning had an account of a new solution of the unification problem which you have recently arrived at and which seems to be a very accurate description of a solution of that problem which was published a year ago by one of my students, Mr. Hoffmann, and myself."[1]

Veblen had only friendly feelings for Einstein, however. In the same letter, he expressed his wish that Einstein visit Fine Hall and see the fireplace with the "subtle but not malicious" epigram. Two years later, Veblen's dream had splendidly come true with Einstein's arrival.

By the time Einstein came to Princeton, Veblen had bid farewell to Hoffmann and had no one there with whom he could work on projective relativity. Although Veblen and Einstein greatly respected each other, and shared an interest in variations of Kaluza's theory, they never took the opportunity to collaborate. Einstein had his own perspective and preferred pursuing his own thoughts at his own pace, with only the help of research assistants whose investigations he could direct. In general, the paucity of collaborations at Princeton was a disappointment for Veblen. Having planned out Fine Hall as a place for scholarly discussion, Veblen regretted he didn't see it happening. "All these mathematicians," he once told Hoffmann, "they meet once a month, then each goes to his little cubby-hole and develops psychotic symptoms almost and has no contact for a whole month with fellow mathematicians."[2]

Meanwhile, Hoffmann, a gifted British-born, Oxford-educated mathematician, had moved on to a position at the University of Rochester. He obtained the position due to sheer circumstance. George Eastman, the founder of Kodak, had just died, leaving millions of dollars to the university. After finding out about the endowment, Hoffmann wrote to the university to see if the money had created an opening. Fortunately, it had, which was lucky for Hoffmann since the job market was tight and he had no other offers.

Before leaving Princeton for Rochester, Hoffmann had written hopefully to Einstein and asked if he could work with him.[3] In the letter, aside from listing his scientific credentials, Hoffmann also mentioned his musical abilities—apparently a great plus for working with perhaps the world's most famous amateur violinist. Although Einstein expressed interest in Hoffmann's work, he was in little position to take on new assistants. Mayer was the only assistant he could support at the time.

At Rochester, Hoffmann studied the issue of how particles move in five dimensions, generalizing Kaluza-Klein theory to bodies with magnetic properties as well as electric charge. He dutifully reported his findings to Einstein in a series of letters. Einstein found Hoffmann's research very interesting and was pleased to collaborate with him after Veblen brought him back to Princeton in 1936.

Assistance, Please

The issue of research assistants continued to be a sore point for Einstein in his relations with Flexner and the institute. Flexner was still upset that Einstein put so much pressure on him to make Mayer's position permanent and independent. Thus he held Einstein responsible for any difficulties that resulted from that agreement.

Once at Princeton, Mayer gravely disappointed Einstein. After writing only one paper with Einstein, he soon lost interest in collaboration. Instead of working on unified field theory and other aspects of relativity, he chose to go off by himself and pursue pure mathematics. Though still at the institute, he was of no use to Einstein. Einstein's plan had backfired—Mayer's position was *too* independent. And because Mayer had tenure, Flexner at first wouldn't let Einstein make another appointment, someone who would genuinely work with him. So Einstein was stuck without a real assistant for a few years.

With the loss of Mayer, Einstein was without his "calculator." He had become used to doling off to Mayer any calculations that needed to be done and spending his time contemplating the big picture. Like an aging architect, he needed robust young workers to complete the constructions he envisioned. Because Flexner wouldn't provide any, Einstein decided to entreat others to help him. Whenever a mathematically talented visitor to the institute expressed interest in his ideas, Einstein charmingly invited them to contribute to his projects. Fortunately, the institute was full of such talent, and Einstein soon could resume his work—albeit with a somewhat different focus for a while.

For a few years, Einstein saw little fertile ground for enlarging upon his goal of unifying the natural forces. Likely, the visitors to the institute didn't stay long enough to offer him sufficient help with such a far-reaching project. Temporarily, he let this grand program lay fallow, and began to till the soil of more mainstream questions.

By returning to basic issues involving standard general relativity and quantum physics he was able to plant the seeds of several incisive ideas, which, through the nurturing influence of his collaborators, blossomed into several significant papers.

Nathan Rosen, a recent Ph.D. from MIT, arrived at Princeton with the intention of studying molecular physics. After discussions with Einstein, however, he became entranced and his aspirations markedly changed. Soon they were working together on deriving general relativistic solutions without mathematical horrors called singularities. As another project, along with the physicist Boris Podolsky, they developed a final rejoinder to quantum uncertainty, a thought-provoking argument called the EPR paradox that caused Bohr's heart to skip a beat. Einstein and Rosen also examined the issue of whether gravity can propagate in waves, and studied how relativistic particles move and interact.

This last topic, called the problem of motion, allured several more institute visitors to come work with Einstein. These included Banesh Hoffmann, who needed little persuasion to begin a long-awaited collaboration, as well as Leopold Infeld, a Polish-born researcher who had known Einstein back in Berlin. Hoffmann recalled how nervous he was on the day he met Einstein:

> I had made some relativistic calculations, and a friend suggested that I go see Einstein to ask his opinion of my work. The idea of my going to see Einstein seemed to me preposterous. I was far too scared. My friend almost had to push me to the door of Einstein's office. I knocked timidly, and Einstein called out the single word "come" with a friendly, rising inflection. I entered in fear and trembling, and there was Einstein sitting in a comfortably chair, sloppily dressed, his hair awry, a pipe in his mouth and a sheaf of calculations on his lap. . . . He smiled and gently asked me to put my equations on the blackboard, and then came these words, which I shall always remember, "Please go slowly. I do not understand things quickly." This from Einstein! At once, as if by magic, all my fears left me.[4]

Einstein gave Hoffmann and Infeld two possible choices of topics, but they were most interested in pursuing the motion question. Together with Einstein, they helped prove that the field equations of general relativity could describe the movements of particles in as direct a manner as Newtonian physics. This placed the theory on even firmer ground.

As fascinating as these problems were, Einstein still hankered to return to the issue of unification. He ardently hoped to have the opportunity at Princeton to resume the process of designing and testing various approaches. Fortunately, he found all the help he needed with two bright young assistants, Peter Bergmann and Valentine Bargmann.

A Mother's Plea

Peter Bergmann was born on March 24, 1915, to royal scientific stock. His father, Max Bergmann, was a rising star in the field of protein chemistry. Working at the University of Berlin in the lab of the Nobel Prize–winning chemist Emil Fischer, the elder Bergmann analyzed the complex structures of long chains of amino acids. Upon Fischer's death in 1919, he assumed a new title and continued this groundbreaking research. Then, in 1921, he was appointed director of the newly founded Kaiser Wilhelm Institute for Leather Research in Dresden, where his work became world famous.

Beginning his career as Einstein's assistant, Peter Bergmann became one of the founders of quantum gravity and one of the leading lights of general relativity.

Peter's mother, Emmy Bergmann, was a well-respected pediatrician and educator. Working at the Empress Auguste Victoria Hospital in Berlin, she developed a special interest in the rights and welfare of children. This advocacy extended to her parental role. Raising Peter and his sister, Esther, she encouraged them to be independent-minded thinkers, and provided whatever assistance they needed for their educational development. She was a strong believer in musical training, and nurtured Peter's talent for playing the violin. Reportedly he had perfect pitch.[5]

Unbeknownst to Peter, wrapped in the warm blanket of youthful innocence, his parents' marriage was strained. Soon after Max assumed his new title, they quietly separated. Emmy moved with the children to the quaint German city of Freiburg in the Black Forest region, taking them regularly to see their father. She did not have the heart to tell them that their father was seeing another woman, the youthful Martha Suter. Max divorced Emmy and married Martha in 1926.

Emmy's older sister, Clara Grunwald, was a close associate of renowned educator Maria Montessori, and brought Montessori's child-centered methods to Germany. Acquiring a great respect for these self-paced learning strategies, Emmy followed in the footsteps of her sister and founded her own school, at her residence in Freiburg. Along with her own children, she admitted more than twenty other pupils, helping them, through mathematical and other educational materials, realize their potential as thinkers.

In this stimulating, self-directed environment Peter acquired the skills to become an independent-minded scientist. It's likely that the mathematical tools he worked with helped him in his later quest to understand the multidimensional geometries of nature. He whizzed through school, graduating when he was only sixteen, and then beginning university. It was then he resolved to become a theoretical physicist.

He was also quite determined in his romantic pursuits. After meeting a young woman, Margot Eisenhardt, who shared his scientific interests, he biked dozens of miles through the Black Forest each time he wanted to see her.[6] His confidence in this domain was right on target. They would get married in 1936 and spend sixty-five happy years together.

All was going well for Peter's career, until it was thrown into turmoil by Hitler's accession to power. Peter was interested in relativity, verboten under the new Aryan creed because of its associations with Einstein. Peter's own ethnic background made matters even worse. As a Jewish boy studying a so-called "Jewish science," he could never have an academic career under the Nazis.

Emmy knew she had to act, to get Peter out of the country and get him a proper graduate education. But where to send him? To get advice on this matter, she decided to write to Einstein. Until the Nazi era, Einstein had traveled in some of the same scientific circles as her ex-husband, such as the Kaiser Wilhelm Society. Perhaps he would be sympathetic to the plight of a bright young physicist in training. Maybe, she hoped, Einstein would even take Peter on as a student.

In her letter, she described Peter as especially talented in mathematics and physics, with also a gift for music. Not knowing of Einstein's plans to go to America, she presumed he would end up in Paris, and imagined Peter studying there under his tutelage. She didn't realize that Einstein never took on graduate students, only fully trained assistants. At the end of the letter, she lamented her family's dire economic conditions, and expressed her hope that her son could have a promising scientific career elsewhere. The letter was smuggled to Le Coq sur Mer, Belgium, where Einstein was residing at the time.

Einstein wrote back an encouraging response, expressing delight that such a talented young student was interested in working with him. He advised Emmy to send Peter to a Ph.D. program outside Germany. As one good option, he suggested that Peter study in Zurich with Wolfgang Pauli. Then, after receiving his Ph.D., Peter would be welcome to join him at the institute in Princeton.

Emmy took Einstein's words seriously. Clearly Pauli would be a great choice as an adviser. Unfortunately, Zurich was exceedingly expensive. Prague, another possibility, was much cheaper. So, mainly for economic reasons, Peter chose to move there and study with Philipp Frank.[7] Frank had replaced Einstein when he had left Prague many years earlier, and later was to write a well-known biography of Einstein. Though certainly not as brilliant and famous as Pauli, he was a respected scholar, and therefore a sound alternative. Frank guided Peter Bergmann along the twin roads of relativity and quantum physics, and helped prepare him for his future work with Einstein.

Valya's Journey

Coincidentally, around the same time, another bright young physics student with a very similar surname unwittingly took the path that Einstein had suggested for Peter Bergmann. Seeing the need to get out of Germany as quickly as possible, Valentine Bargmann, a Berlin University student of Russian Jewish ancestry, decided to flee to Switzerland and continue his studies at the University of Zurich. Though Valya—as his friends called him—obtained his Ph.D. under the guidance of physicist Gregor Wentzel, his mathematical talents were keenly noticed by Pauli. Consequently, instead of getting to know a Bergmann, Pauli became well acquainted with a Bargmann.

Born in Berlin on April 6, 1908, Bargmann was a quiet, hard-working, well-organized student with a proclivity for mathematics. A talented pianist, he enjoyed performing solo or accompanying other musicians during social gatherings.[8] This would prove an essential skill while working with Einstein.

Bargmann came to the Institute of Advanced Study through a roundabout course. When in 1936 he received his degree, he couldn't remain in Switzerland because the country was not accepting immigrants. Instead he joined his parents, who had fled to Lithuania. Fortunately, his parents knew a secretary at the American Consulate who obtained a visa for him. Bargmann emigrated just in the nick of time; his passport would have expired two days later.

Once in the United States, not knowing where to go, Bargmann wrote to Pauli for advice. Pauli steered him to Ann Arbor. Once he arrived there, however, he discovered that there were no funds available. Consequently he was redirected to Princeton. There he met the famous mathematician John von Neumann, who appointed him as a nonsalaried member of the institute.

In due course, Bargmann found out about Einstein's work and became interested in participating. Hermann Weyl, who had left Göttingen for Princeton because of the Nazi terrors, introduced him to Einstein. Soon Einstein warmly invited him to collaborate as an unofficial assistant.

Family Struggles

In 1937, when Valya Bargmann arrived at the institute, Peter Bergmann had already been working there for a year. When he was about to get his degree he had written to Einstein from Prague, citing his father's name as a reference. The letter mentioned his great interest in uniting relativity with quantum mechanics.[9] At first, to Peter's disappointment, Einstein did not reply. After a month of waiting, Peter wrote again. This time he received a friendly response, cordially inviting him to become his official assistant. Thus Peter's appointment had been more formal than Valya's.

Peter later found out the reason for the delay. In the interim, Einstein wrote to Philipp Frank enquiring about his credentials. After Frank wrote back a glowing letter of reference, Einstein felt very comfortable bringing Peter to Princeton.

By that point, Peter's father had emigrated to America as well. Max Bergmann, like many other German Jewish scientists, had been forced to "retire" from his position in Dresden. Fortunately, the illustrious physician Simon Flexner, Abraham's well-known brother, had recognized Max's talents and offered him a prestigious position at the Rockefeller Institute for Medical Research (later Rockefeller University) in New York. Coincidentally, Simon Flexner headed the institute where Max Bergmann worked, just as Abraham Flexner directed the institute where Peter assumed his assistantship.

Max Bergmann was well respected at Rockefeller, and helped attract talented biochemists to his department, including future Nobel Prize winners Stanford Moore and William H. Stein. The Bergmann Laboratory won international renown for its research on the properties of proteins and enzymes. Max took special pride in assisting many other emigrant scientists with finding jobs in the United States, writing numerous letters to the Emergency Committee of Displaced Foreign Scholars.

Once Einstein received a letter from a petroleum company that needed a chemist. Not knowing anyone himself, he passed on the information to Max Bergmann, who immediately found the name of an Italian immigrant chemist who had fled because of racial laws. Cheerfully supplying Einstein with the name, Max took the opportunity to express deep gratitude for his help with Peter's career.[10]

Peter's mother, who stayed in Germany until the time of World War II, wasn't so lucky with her career. Under the Nazis, child-centered educational programs were discontinued. Seeing her pedagogical achievements crumble around her, she fled to Palestine in 1939. Her better-known sister, Clara Grunwald, remained behind and perished at Auschwitz.

The Berg and the Barg

When Bargmann joined Bergmann to be his two closest assistants, Einstein was living in a white-shingled house on Mercer Street, conveniently situated down the road from Princeton University. It was a modest house, considering its famous occupant, smaller than many others on the street. Einstein enjoyed walking or bicycling from there to his office, sometimes stopping for ice cream on the way back.

Einstein lived and worked at this house in Princeton during the final decades of his life.

Elsa had died in December 1936, little more than a year after they had moved in. With her passing, Einstein focused even more on his research, shunning even the scattered social engagements he used to allow himself. The petty details of his life became the exclusive domain of Helen Dukas, who also lived at the house. Upon Elsa's death, she had assumed the roles of housekeeper, general manager, and confidante as well as personal secretary. Einstein's stepdaughter Margot also resided with them.

Dukas had a ready sense of humor. Amused that Einstein's two assistants had such similar names, she affectionately nicknamed them the Berg and the Barg.[11] This became a running joke for many years.

Down the road from Einstein lived the physicist John Wheeler, his wife, Janette, and their children. Wheeler, a student of Herzfeld and a close collaborator with Bohr, was a pioneer in models of nuclear structure. Although general relativity was far from his training, he greatly admired Einstein and would become one of the great interpreters and explicators of gravitational theory.

Wheeler recalled seeing Einstein walking to and from the institute, always with Bergmann and Bargmann: "I remember so often seeing them together. I always had the feeling they were smaller than Einstein. He dominated the picture. One day Einstein called up to say, 'Your children's cat is over at my house.' It followed the three of them from the Institute for Advanced Study back to his house. I asked the cat if it learned anything."[12]

Einstein and his assistants fell into a steady working routine. Every morning they would meet at Einstein's office to discuss ideas. They would show him their calculations and ask for advice on how to proceed further. Even though he performed none of the equation shuffling himself, Einstein would make detailed, incisive comments—sometimes encouraging his assistants to continue, other times to switch abruptly to a new direction. Whenever they were discouraged by the difficulty of their task and felt as if they were bashing their heads against a brick wall, he would comfort them with encouraging words. "The world has waited this long," he would tell them. "Another few months won't make much difference."[13]

At lunchtime, Einstein would leave Fine Hall and return home. Though he would usually spend his afternoons in solitary thought, he made himself available for consultation, if the situation arose. Sometimes his assistants would call him at home; other times they would just drop by. Otherwise, they'd be immersed in the details of calculation, readying themselves for their next day's meeting with their mentor.

Einstein's assistants were amazed by his dogged persistence. After discarding one version of a theory, he would waste no time before proposing another. As Bergmann remarked: "Whenever Einstein was working on a new model of field theory, he would be tremendously enthusiastic for weeks and months, and even for years, but inevitably the day of reckoning would come when Einstein would be the first to find a fatal flaw. With complete ruthlessness, he would discontinue this theoretical approach and come up with a completely new idea, usually within days."[14]

Often, music would serve to replenish his spirit. When Einstein fled Europe, his piano was one of the cherished possessions he managed to keep. It dominated one of the first-floor rooms of his house. Naturally, he held onto his beloved violin as well, which he would take out regularly for practice. The strains of beautiful melodies would gloriously emerge from the physicist's hands, releasing the tensions of equation writing. "He loved to perform,"[15] recalled Bergmann, who enjoyed the spontaneous recitals.

With a repertoire that included Mozart, Bach, Schubert, and Vivaldi, Einstein would invite Bargmann and others over at least once a week to play chamber music. Bargmann found it "extremely gratifying to make music with him."[16] Then, refreshed by the musical interlude, they could resume their work in high spirits.

Sailing provided Einstein with another means of relaxation. While drifting through Long Island Sound on his little boat, the *Tinef* (roughly meaning "worthless"), he found some of his greatest pleasure. With his long hair tossed by the wind as the waves crashed around his craft, he resembled a maritime Stokowski conducting a soaring Beethoven opus. Once the air grew still and the sails needed no more adjustment, he would enjoy a calmness of mind as deep and timeless as the sea. Then his thoughts could wander through the oceans of space and time, dreaming about a grand mathematical symphony that could express the beauty around him. Leaving salt and spray behind, Einstein would return to Princeton with even more determination to discover the score for nature's universal composition.

Thou Shalt Not Sin

With Bergmann and Bargmann, Einstein revisited the Kaluza-Klein idea for his last and most comprehensive attempt at higher-dimensional unification. Systematically, they struggled to combine gravitation and electromagnetism into a single theory that would extend Einstein's standard equations of general relativity. During this venture Einstein was far more open than on other attempts to consider more far-reaching alternatives, including the possibility of a physical extra dimension.

As he guided his assistants during their exhibitions into un-known territory, it became clear to them that he had very fixed ideas about what features should or should not become part of a unified field theory. Despite his disinclination to mix science with religion (in the conventional sense), these injunctions took on a biblical tone. Einstein based the legitimacy of a "Theory of Everything" on whether or not God would have made the universe that way. In that sense, his guidance was an attempt to read and interpret divine preferences. "Let me see, if I were God which one of these would I choose?"[17] he would sometimes remark when considering various options.

Like a minister counseling his flock, Einstein presented his assis-tants with an implied list of virtues and sins. Knowing his taste, they would strive hard to make their models more "virtuous" and less "sin-ful." One of the cardinal sins, for example, was bringing any notion of probability into the theories. Anything resembling quantum jumps was explicitly forbidden. Consequently, even though Einstein recognized Klein's contributions, he consistently would not replicate Klein's linkage between the fifth dimension and quantization.

One time, later in their careers, Klein visited Princeton and took the opportunity to ask Einstein what he thought of the theory of sec-ond quantization. Einstein replied, "Second quantization. That is sinning squared."[18] Philosophically, he would have preferred that Klein had chosen the causal path instead of siding with Bohr.

Another of Einstein's prohibitions was treating gravitational and nongravitational fields in any way differently from one another. Einstein felt that standard general relativity made an artificial distinction between gravity, described geometrically on the left-hand side of his equations, and electromagnetism, included almost as an afterthought on the right-hand side when circumstances required it. He wanted to see all forces delineated in like manner on the left-hand side, with nothing on the right-hand side. For similar reasons, he also preferred mass and energy to have geometric descriptions so that they too could reside on the left-hand side of the equations.

There were mathematical as well as physical transgressions. Einstein urged his assistants to avoid shortcuts that lacked mathematical rigor. With his research viewed through dubious eyes by most physicists, he could little afford even the hint of an inaccuracy. "You must not sin,"[19] he would insist if they even suggested such a shortcut.

Finally, one of the deadliest sins was constructing a theory with singularities or unspecified free parameters. Singularities, as regions where quantities become infinite, represent breaches in the laws of physics. Unfettered parameters, allowing an unlimited variety of possibilities, similarly reject physical experience. There should be only one possible universe determined by the ultimate set of equations, not many. Translated into Einstein's scientific "theology," these restrictions meant that God's construction of universal laws must have been complete (no loose ends, such as singularities) and unique (no other solutions).

Virtuous unification theories, in contrast, included those that were simple and comprehensive, and duplicated known experimental results, such as the bending of light and the procession of Mercury's orbit. They were also to be invariant under the same transformations obeyed by ordinary general relativity—maintaining familiar properties such as the equivalence principle. As in general relativity, someone in a windowless spaceship should not be able to tell the difference between plunging freely toward Earth versus resting comfortably in deep space.

Lastly, grounds for sainthood in the Einstein canon would be the creation of a theory that reproduces atomic spectra and other quantum features *without using probability*. This would be the ultimate way of reconstructing atomic theory within a purely causal framework.

One of the means by which Einstein hoped to achieve this was by use of a mathematical circumstance known as *overdetermination.*

Ordinarily, in solving a problem, one expects the number of equations to match the number of unknowns. For example, in the case where Johnny is twice the age of Jenny and he is six years older than she is, there are exactly two variables (their ages) and two expressions relating them. An overdetermined set of equations, on the other hand, has more equations than variables. Normally, this greatly restricts the results—that is, if a solution can be found at all. Einstein hoped that such a situation would produce a discrete spectrum of solutions, yielding "quantum conditions analogous to the Bohr orbits."[20]

From Einstein's perspective, the introduction of a fifth dimension presented many opportunities. It augmented standard general relativity with additional field equations, possibly resulting in overdetermination. The added metric terms could be viewed as extra fields, conceivably replacing the quantum wave function. This could lead, perhaps, to a fully causal substitute for quantum mechanics. Possibly the fifth dimension could also explain seemingly nonlocal effects, replacing them with unseen connections. As Bergmann has written, "since the description of a five-dimensional world would be incomplete, it was hoped that the indeterminacy of 'four-dimensional' laws would account for the indeterminacy relation and that quantum phenomena would, after all, be explained by a [classical] field theory."[21]

In short, one of the principal reasons Einstein embraced the fifth dimension during that time in his career was the hope that it would add bizarre complications to general relativity, mimicking the strange quantum rules. But unlike the latter, the former would fundamentally be completely deterministic—predictable through the five-dimensional extension of his equations.

Getting Physical

Einstein realized by that point that the utility of the fifth dimension depended on it being physically real. Based on his experience with Mayer, he had concluded that a purely mathematical extra dimension would not achieve his dual purpose of unifying the natural forces while undercutting quantum theory. As he wrote in a 1938 paper coauthored with Bergmann:

> If Kaluza's attempt is a real step forward, then it is because of the introduction of the five-dimensional space. There have been many attempts to retain the essential formal results obtained by Kaluza without sacrificing the four-dimensional character of the physical space. This shows distinctly how vividly our physical intuition resists the introduction of the fifth dimension. But by considering and comparing all these attempts one must come to the conclusion that all these endeavors did not improve the situation. It seems impossible to formulate Kaluza's ideas in a simple way without introducing the fifth dimension.
>
> We have, therefore, to take the fifth dimension seriously although we are not encouraged to do so by plain experience.[22]

In that paper, as well as in a second article written by Einstein, Bergmann, and Bargmann, the fifth dimension enters the equations as an equal among its brethren. Nothing distinguishes it, at first, from space and time. This state of freedom does not exist for long, however. The authors impose a periodicity condition, mandating that it effectively be rolled up into a tube.

Wasn't that Klein's portrayal of the fifth dimension? Pauli pointed out the obvious similarities to Einstein, after he heard about the theory from Bargmann. "That is indeed an old idea of O. Klein,"[23] Pauli wrote.

Einstein insisted to Pauli, however, that "the new work has only a superficial similarity to Klein's. It is a logical improvement of Kaluza's idea, that deserves to be taken and examined seriously."[24]

The major difference between the two theories is that while Klein's relates the size of the fifth dimension to the quantization of electric charge, Einstein, Bergmann, and Bargmann make no such connection. In fact, they make no mention of quantum theory at all in their articles, and only briefly in their correspondence to each other. Scholar Jeroen van Dongen finds this omission a real puzzlement, perhaps an implicit recognition of their failure to mimic quantum theory.[25]

One major problem they do indeed point out in the second article is their model's prediction that the electromagnetic and gravitational interactions should be of equal strengths. In nature, electromagnetism is far stronger than gravity, a fact that the theory doesn't bear out. They attribute the difficulty to the "fact that the equations are uniquely determined . . . [with] no arbitrary constants."[26] What Einstein considered a blessing—unique solutions with no free parameters—turned out to be a curse, because they couldn't make any adjustments to model nature more closely. If they did have a free parameter, perhaps they could have adjusted it to yield more natural results.

Anyone familiar with the physics of the times would have noted another feature lacking in their approach. Their unification model included only two forces, yet by then other interactions were known. One of the most pressing issues of the day concerned the properties of the forces that held nuclei together and allowed them to decay, namely what we now call the strong and weak interactions. It was the talk of physics, but Einstein was so secluded that he wasn't really listening.

Nuclear Matters

When Bohr visited Princeton for four months in 1939, he and Einstein exchanged pleasantries but little more. They did not vie with each other in fiery debate, nor did they embrace each other in tearful reconciliation. In fact, each was so deep in thought that they barely interacted with each other at all. Einstein was so enmeshed in the details of unification that he had little else to say. He gave one lecture on the subject during that time, which Bohr attended out of respect. At the end of the talk, he cast steely eyes on Bohr and emphasized that he hoped his theory would reproduce the quantum rules. Bohr remained silent.[27]

Bohr, on the other hand, was solemnly mulling over the implications of the recent discovery of nuclear fission by German scientist Otto Hahn. In 1938, Hahn bombarded uranium nuclei with neutrons, splitting some of them into barium nuclei and producing energy in the process. Finding out about this earth-shattering discovery, Bohr, collaborating with Wheeler, wished to use his time at Princeton to explore the theory behind these results.

The discovery of nuclear fission proved the culmination of a decade centered on the minute centers of atoms. Just as the outer wrappings of the atom were the prime target of study for quantum physicists in the 1920s, its inner core was the major focus for nuclear physicists in the 1930s. Through studying the process of nuclear radiation, they managed to illuminate what they had once considered a black box.

Scientists had known since Henri Becquerel's 1896 discovery that atoms release particles through the process of radioactive decay. There were three observed types of decay. Beta decay seemed to produce electrons. Alpha decay, on the other hand, appeared to generate heavier, positively charged particles, and gamma decay seemed to involve invisible, energetic electromagnetic radiation.

In 1930, Pauli applied physical principles to beta decay, concluding from missing energy that hitherto undetected particles must be produced at the same time as electrons. The unseen escapee, he surmised, would be electrically neutral. Two years later, James Chadwick did indeed discover a neutral particle emanating from atomic nuclei, but it was far too massive to be Pauli's interloper. The detected object was dubbed the neutron, with the invisible particle assuming the moniker neutrino, so named by physicist Enrico Fermi. Fermi reinterpreted beta decay as the spontaneous disintegration of neutrons into protons, electrons, and neutrinos. However, this process, an example of the weak interaction, would not be fully understood until decades later.

With the discovery of the neutron, physicists were hard pressed to provide an explanation of how atomic nuclei hold together. Electromagnetic repulsion would rapidly blast positively charged protons apart, leaving neutrons in their wake as innocent bystanders, if it weren't for an unknown force sealing them tight. Except for hydrogen, the lightest element, no atomic nuclei would be stable.

To account for the stability of heavier nuclei, physicists postulated the existence of a strong nuclear interaction that would kick in at miniscule scales. The strength of this interaction would drop off with distance so rapidly that it could never be detected beyond the close confines of nuclei. Moreover, it would only affect protons and neutrons, among the particles then known, leaving electrons (and neutrinos) free to wander off.

In attempts to explain the workings of the strong interaction, Heisenberg proposed a mechanism involving the exchange of particles. He imagined protons and neutrons volleying electrons from one to another like children playing catch. Because each would want its electron back, the process of sharing would force them to stick together. Moreover the symmetry between the proton and neutron, as expressed in their similar masses and ability to transform from one into the other, would group them into a "doublet." A new quantum number, called *isospin* (also known as isotopic spin), would distinguish the two members of the doublet. The proton would have "up" isospin and the neutron "down" isospin. Heisenberg's theory violated a special conservation law, however, because protons and neutrons, though with different isospin, each have the same ordinary spin. If they shared an electron that also had spin, they'd be creating extra spin out of thin air.

In 1935, the Japanese physicist Hideki Yukawa vastly improved upon Heisenberg's model by replacing the electron with another exchange particle. He proposed that a new particle two hundred times as massive as the electron would be better suited to do the job. The meson, as this exchange particle would eventually be called, could be found in differently charged varieties, allowing for various kinds of interplay. Bohr was skeptical about this idea at first, accusing Yukawa of inventing particles just for fun.

Then, in 1937, nature played a sly prank on the physics community. That year, several teams of experimentalists detected a particle amidst cosmic radiation that seemed to have the correct mass to be Yukawa's exchange mechanism. This discovery vaulted Yukawa's theory to near-universal recognition as the proper explanation of the strong nuclear force. Bizarrely, the particle detected that year, called the muon, had nothing to do with the nucleus. It was pure coincidence that it was found at that time. The gradual realization of this prompted physicist I. L. Rabi to exclaim, "Who ordered that?"[28] A decade later came the discovery of the first true Yukawa meson, called the pion, bringing closure to the Japanese physicist's quest.

Anyone keeping up with these developments would have attempted to include the ever-growing array of particles and forces in any unified field theory. If they couldn't do so, then they would at least have recognized that their model was far from complete. That's why it was eminently clear to Bohr, Heisenberg, Wheeler, and others studying nuclear structure that Einstein was completely out of touch.

Even Bergmann, more attune with the quantum community than Einstein, keenly noticed this omission. In August 1938, while vacationing in Maine, he wrote Einstein with profound concerns that the five-dimensional theory did not seem to encompass neutrons.[29] Einstein, in response to the new discoveries, would pay little more than lip service at most. Like a Victorian cottage surrounded by high-rise developments, by that point in his life he preferred traditional aesthetics to newfangled innovations. He became interested in nuclear theory only when its implications for global politics became all too apparent—famously warning Roosevelt about the possibility of a Nazi bomb. But this interest did not extend to a reworking of his own physical concepts.

If Einstein couldn't do the job, then who would suggest the means for natural unification—including the nuclear forces as well as electromagnetism and gravity? Amazingly Oskar Klein, who once abandoned the fifth dimension like a confirmed bachelor, would take up again with his old flame. He would present the first known proposal for full unification in higher dimensions at a 1938 spring conference in Warsaw, Poland.

Klein's theory, though ultimately unsuccessful, contained premonitions of many future developments in physics. If it were more widely known, it might have hastened certain modern conceptions of particle interactions. Though published in the conference proceedings the following year, the article was barely read, overshadowed instead by news of war. Like his jaundice putting him on the sidelines during the heyday of quantum theory, it was yet another case where Klein's good intentions were thwarted by bad timing. With Hitler's invasions spurring the outbreak of World War II, it was a bitter age for Europe and unity of any sort seemed a distant dream.

CHAPTER 9

Brave New World

Seeking Unity in an Age of Conflict

There are two ways to respond to tragedy. One of them is to give in, become despondent—and one or two [physicists] were. The other is to get to work. They were working on two fronts. One was their science, almost in the style of Brave New World where there's soma. Physics was soma for some of them; one of them even said so. The other was that they were politically very active, trying to tell people what was going on in Europe.

—GERALD HOLTON, scientific historian

Twilight of the Old Europe

Stockholm has a pleasanter climate than its latitude would suggest. Warmed by a side branch of the Gulf Stream, it usually manages to escape the worst of the Arctic weather. Though north of Labrador, its winters are more tolerable, its port welcoming ships year-round.

The 1930s found Oskar Klein quite happy in his own cozy harbor. He and Gerda took pride in raising their large family. He marveled at his children's diverse talents, ranging from the humanities to biology. They would end up pursuing art, teaching, journalism, and other professions, but not physics. Although Klein would have loved to have seen one of his children take an interest in his own profession, he was nevertheless very open-minded and accepting of whatever path they wished to follow.[1]

Content as he was, the Swedish Odysseus could not remain landlocked forever. The siren song of the fifth dimension was too sweet and compelling. Try as he might, he could not vanquish its tune from his mind. He needed to set sail once more.

In 1938, Klein crossed the Baltic to central Europe. Though bathed in the same temperate waters as Sweden, the atmosphere there was much more frigid and frightening. Politics, rather than geology, had caused this climatic shift. Hitler's detestable policies were plunging international cooperation into an ice age, taking the European physics community with it as one of the casualties.

Attendance at the Warsaw conference, "New Theories of Physics," reflected a world that was sharply divided. The conference was jointly sponsored by the International Union of Physics and the Polish Intellectual Cooperation Committee, groups promoted by the League of Nations for peaceful discourse among scientists. With Hitler's designs upon Poland—which he would savagely invade the following year—such cooperation ran contrary to his ultranationalistic views. Consequently, Heisenberg and other German scientists were prohibited from attending. For similar reasons, no Italian scientists were in attendance either. Nevertheless, a number of prominent physicists from other countries managed to participate, including Niels Bohr, Arthur Eddington, George Gamow, Hendrik Kramers, Léon Rosenfeld, and Eugene Wigner.

Klein's Tour-de-Force

To little fanfare, Klein presented his new five-dimensional work, "On the Theory of Charged Fields." It offered an intriguing way to combine gravitational, electromagnetic, and nuclear forces in a unified approach based on Dirac's successful model of the electron. Unlike the models proposed by Einstein, Bergmann, and Bargmann, it incorporated Yukawa's discovery and other recent innovations in particle physics. In particular, it allowed for protons and neutrons to transform from one into the other by exchanging mesons (which he referred to by an earlier designation, mesotrons). This corresponded to an essential symmetry between protons and neutrons, as if they were near-twins.

Mathematically, Klein framed his model in terms of a two-row matrix. One of the rows corresponded to zero charge, housing the neutron field, and the other to positive charge, encapsulating the proton field. As in Yukawa's theory, the fields formed an isospin doublet. Through clever machinations, the different charges of each member of the doublet arose from its mathematical dependence on the fifth dimension. Another matrix, representing the meson, permitted transformations from one field to another. These constituted rotations in a kind of abstract domain called isospace. A parallel structure described similar transformations between electrons and neutrinos. Klein's theory characterized the weak and strong forces by means of a single mechanism. We now know that they operate distinctly, with separate types of exchange particles.

One can envision Klein's model in terms of compass directions, with the proton field representing "north" and the neutron field representing "south" in isospace. In general, nucleons (protons and neutrons) are a combination of the two fields, depicted as an intermediate direction—"northeast," for example. Mesons act to rotate nucleons from one orientation to another, changing their mixture of states. The modern designation for this state of affairs is that nucleons experience a gauge symmetry, with the exchange particle serving as a gauge field. This is a different kind of gauge model than what Weyl originally described, operating in an abstract domain rather than in actual space-time.

Published in French as part of the conference proceedings, Klein's theory was unearthed more than a decade later by physicist (and future son-in-law) Stanley Deser. Deser describes coming across the paper:

> I was a beginning graduate student at Harvard and we were trying to soak up physics in various ways. I remember one day wandering up there in the stacks of the fourth floor physics library and found this unbelievably dusty old text. I came to the US from Europe so French happened to be a language I was familiar with, and I saw this dusty volume and thought "who were all these guys?" I didn't know much physics. But then I saw Klein's paper and thought, "gee

that sounds pretty good. I wonder what happened. Where is he now? What happened to all this physics?"[2]

Deser would later show the article to his adviser, Julian Schwinger, and then it would become part of the folklore. Physicists would marvel at its premonitions, while noting one of its major lackings. Successful gauge theories of nuclear interactions, pioneered by Chen Ning (Frank) Yang and Robert Mills in 1954 (with no knowledge of Klein's work), would be *non-Abelian,* meaning that the order of transformations made a profound difference. Rotation A preceding rotation B would produce a different result than if the succession were reversed. A special group of gauge transformations, called SU(2), embodied this property. In mathematics, a group is a set of elements and operations governed by particular rules. In the case of SU(2), these elements are two-by-two matrices with unique features. For the symmetries (invariances under transformation) of this particular group to manifest themselves, the gauge particle needs to be massless.

One way of representing the SU(2) group is by means of rotations along the surface of a sphere. As illustrated by the case of Marius and Darius walking along separate paths to the North Pole (described in chapter 4), the order of latitudinal and longitudinal movements profoundly matters. North followed by east produces sharply different results than east followed by north. Such is the reason that mathematicians classify SU(2) as a non-Abelian group.

Klein's model did not involve SU(2) at all. Rather it was based on an extension of U(1), the simpler group of electromagnetic gauge transformations, with accommodations for nuclear forces tacked on. U(1) can be represented by rotations around the circumference of a flat circle. For such transformations, the order of operations does not matter. Rotating 45 degrees clockwise and then 30 degrees counterclockwise along the edge of a ring, for example, yields the same result if these operations are reversed.

Also in contrast to the later Yang-Mills theory, Klein's gauge particle, the meson, was massive. Thus he almost created a modern gauge theory, but not quite. Still, physicists of the 1950s and thereafter would be impressed by how close Klein's theory came to the mark considering how early it had appeared. His intuitions had somehow lept more than a decade and a half ahead, without the experimental evidence or theoretical machinery required to justify his bold assertions. In his Nobel Prize acceptance speech, Abdus Salam would call Klein's 1938 proposal "a real tour-de-force."[3]

Klein's contemporaries, in contrast, seemed to have totally ignored his work. At the conference, only one physicist, Christian Moller, asked a question. He wondered if the theory could accommodate neutral as well as charged mesons. On the spot, Klein cleverly reworked his theory to accommodate Moller's remark.[4] The rest of the attendees sat in complete silence, unaware that they were witnessing the very future of theoretical particle physics.

Later, Klein wrote to Bohr, inquiring if his theory could be published in *Physical Review,* the leading American journal. Bohr apparently didn't even respond. With Hitler's frightening machinations and a growing refugee crisis—not to mention the discovery of nuclear fission—everyone had more imminent matters to ponder.

From the start, Klein himself was deeply involved in helping scientists flee from Nazi Germany. In 1933, he supplied funds enabling Walter Gordon to escape from Hamburg and resettle in Stockholm. Five years later, he supported Lise Meitner, a former assistant to Otto Hahn who similarly fled to Sweden. Meitner subsequently revealed Hahn's fission discovery to Bohr and other nuclear physicists. Then, once World War II broke out with the invasion of Poland, Klein and his wife acted as a nucleus for refugees in Sweden. It was a dangerous task, but Klein felt a moral duty to do his part—a lingering effect of his father's humanitarian influence, no doubt. Klein's daughter Elsbeth remembers a continuous stream of Danish refugees staying at their house, most of whom were not physicists and had no connection to their family.[5] Most notably, Klein helped Niels Bohr flee to Sweden during his perilous escape from the Nazis.

The Göttingen Purge

While Klein was helping immigrants find refuge, Kaluza was helping a German university function without them. The steady exodus of Jewish and otherwise persecuted professors had created a genuine shortage of experienced faculty. Universities scrambled to maintain their research programs and keep some measure of their previous standing.

Kaluza filled a vacancy opened by the Göttingen purge of 1933. In April 7 of that year, the Nazi government issued a decree authorizing the dismissal of non-Aryan professors unless they had served in World War I. (Later, even veterans would lose their jobs as well.) Soon hundreds of Jewish professors around the country found themselves on forced "leave," including Walter Gordon, Max Bergmann, and many others. Even department heads were dismissed, including Max Born, longtime director of Göttingen's physics department. Born's associate, Nobel Prize winner James Franck, resigned in protest, but the authorities didn't care.

In Göttingen's stellar math department, the effects were particularly dire. A new mathematics institute had recently been established—with a brand-new building just beyond the old town walls—headed by respected mathematician Richard Courant. Courant had been instrumental in securing funding for the center. Nevertheless, the anti-Semitic law forced him out, along with many other colleagues. Emmy Noether, one of Germany's first woman mathematicians, who struggled hard to obtain professorial status, also had to leave. She is remembered today for her finding that symmetries are related to conserved quantities, a discovery that helped pave the way for modern gauge theories. And the inventor of gauge theories himself, Hermann Weyl, soon left Göttingen. Weyl wasn't Jewish, but his wife was. Because of the large anti-Semitic student movement on campus, he feared she would constantly be harassed. Consequently he resigned and moved to Princeton, where he joined Einstein at the Institute for Advanced Study.

Petitions addressed to the government protesting these dismissals were signed by many prominent Germans, including Hilbert and Heisenberg. These heartfelt pleas were met with silence. Einstein wrote around that time to the president of the Bavarian Academy of Sciences that "the learned societies of Germany have, to the best of my knowledge, stood by and said nothing while a not inconsiderable proportion of German scholars and students and also of academically trained professionals have been deprived of all chance of getting employment or earning a living in Germany."[6]

Later Hilbert would be invited to a dinner that included the infamous Nazi education minister Bernhard Rust. When Rust asked Hilbert if the banishment of Göttingen's Jews diminished in any way the mathematics program at Göttingen, he bitterly replied, "There is no mathematics left at Göttingen."[7]

Nevertheless, someone needed to rule over the ruins. With Courant on the way to New York, where he would successfully reestablish his career abroad, Göttingen's math department needed a new director. Briefly, Weyl had taken on the role. After he decided to leave for the United States as well, mathematician Helmut Hasse reluctantly assumed the directorship.

Under Rust, the Nazis enacted a program to "Germanize" science and mathematics, removing so-called "Jewish influences." In physics, anything having to do with Einstein's theories or even quantum mechanics officially became taboo. Paul Ziegenbein, a mathematician from Kiel who studied under Kaluza, was designated to assist Hasse and help administer the "Germanization."[8] As a loyal member of the party, the Nazis entrusted Ziegenbein to keep watch over Hasse.

Though Hasse had assumed Courant's administrative responsibilities, he still needed someone to fill Courant's research and teaching roles. Therefore in 1935 the department conducted a search for suitable candidates. Ziegenbein and Hasse's familiarity with Kaluza's work placed his name high on the list. Ultimately, after inquiring about a few other candidates, they offered him the appointment.

As biographer Daniela Wünsch points out, the choice of Kaluza was surprising in many ways. First, his field was very different than Courant's. While Courant was an applied mathematician, Kaluza made his name in five-dimensional theory and other generalizations of relativity. Kaluza absolutely insisted, before agreeing to the position, that he was not a "second Courant."[9]

Second, Kaluza was never a member of the Nazi party, nor any other National Socialist organization, and had absolutely no interest in joining. He grew up in a liberal tradition, and found racial hatred appalling. He had known Courant and other fired scientists, and was incensed by their dismissal. Nevertheless, Kaluza's principled stance somehow did not preclude his appointment.

Wishing to leave Germany during those dark times, Kaluza would have welcomed a position abroad. Given the job shortage, however, and his lack of research productivity, he felt that he had little chance of getting such employment. Therefore with some trepidation he accepted the Göttingen offer.

Kaluza's Cocoon

By the time Kaluza arrived in Göttingen, the climate had become slightly less political. Two years of extreme "Germanization" had worn out many of the more traditional academics. Therefore Kaluza's appointment fit in well with Hasse's position to steer the department back toward a more mainstream course.

Not that the department or the university became in any way liberal. It was still Nazi Germany. Like any other administrator of the time, Hasse worked hard to maintain favorable ties with the regime. When it was strategically necessary, he even applied for party membership. But, as far as mathematics itself was concerned, Hasse managed to keep politics largely out of it.

With Germany immersed in barbarism and chaos, Kaluza found Göttingen's Mathematics Institute a cocoon in which he could shelter himself. He was given an office on the second floor, down the hall from the department's comfortable library. As his daughter recalled, he spent considerable time sitting in the library and reading. Not only did he peruse mathematics journals, he also kept up with the latest findings in physics, chemistry, biology, and psychology. He was even interested in theories of the paranormal, a popular subject of the times. Dorothea was familiar with her father's practices because she worked at the Physics Institute next door and often visited him.[10]

Kaluza's research at the time fell into two categories. The first was his formal, published work that included an influential textbook, *Higher Mathematics for the Practical Reader*,[11] coauthored with physicist Georg Joos. Into the second category fell his musings, daydreams, and informal projects. These, by far, took up the bulk of his time.

Every day when Kaluza walked through the lobby of the Mathematics Institute, he passed curious displays of geometric models.

Some of these were designed by mathematician Victor Schlegel, who, like Irving Stringham, created realistic projections of hypercubes and other higher-dimensional objects. Felix Klein, who had procured these for the department, considered them excellent tools for instruction. They were ways of envisioning the strange properties of hyperspace.

In his scientific isolation, Kaluza became obsessed with the idea of proving his five-dimensional theory through visual analogy. In the manner of Plato's cave allegory, he wanted to show that the world of the senses is but the shadow of a true five-dimensional reality. He built a special geometric apparatus, similar to those in the display cases, that he could hold up to the light and manipulate, demonstrating that real world forms resemble the projected features of hyperspace.[12] Considering his ruminations on this subject too speculative, he never tried to publish them.

Displayed in the main hall of Göttingen's Mathematical Institute are glass cases filled with all manner of geometric shapes, including projections of hypercubes and other higher-dimensional objects. Undoubtedly, these attracted the interest of Theodor Kaluza, who became a professor at the institute several years after its founding.

Once the war started, Kaluza soon had more pressing concerns. Hasse was appointed to a government position in Berlin, leaving Kaluza as one of only two senior mathematicians at Göttingen. Appointed to the directorship, Kaluza now had the unpleasant job of dealing with the university officials. Nevertheless, he somehow managed to remain independent in his views, epitomized by the fact that he bucked official policy and refused to sign his correspondence with the salutation "Heil Hitler."[13] This disassociation with the Nazis would help him, after they were finally defeated, to maintain his position at Göttingen.

Physics during Wartime

During the first years of the war, as Hitler's regime subsumed ever-growing portions of the European continent, refugees knocked louder upon America's door, seeking shelter from their grave peril. Of those who made it in, many hoped to find positions comparable to the ones they left behind. They competed with those who had arrived earlier, in the prewar wave of immigration, for jobs that were increasingly scarce.

Academics from prestigious European universities, hoping to get new appointments right away at Harvard or Princeton, were sometimes gravely disappointed. Because of the shortage, they often had to settle for places they had never heard of, in the great American hinterland. A smorgasbord of resettlement agencies helped match prospective faculty with openings at small colleges. As Gerald Holton relates:

> A story that typifies this was Victor Weisskopf, a very good physicist coming from Europe with his Danish wife, and trying to find a job at one of the physics conferences—a job fair. He came home and he said, "Ellen I have a job." She said, "where?" "I don't really remember. I think the university starts with an 'R.' "
>
> It was demoralizing. Some of them took it very hard. But the majority did not. The majority rose. MIT got Victor Weisskopf very quickly. Harvard took Philipp Frank.[14]

In this atmosphere, Einstein strived hard to get his assistants positions at major American universities. When Bergmann's position at the institute expired in 1940, Einstein sent out letters of reference to a number of key places. Through Richard Tolman, Einstein learned about an opening at Caltech, and dashed off a glowing recommendation: "I have had for four years a young assistant, Dr. Peter G. Bergmann. We have worked together daily so that I have been able to form a sound judgement of him. He has solid mathematical knowledge, power of clear and independent thought and of constructive formulation. He was a very valuable help to me and I greatly esteem him not only on account of his work but also for his upright character."[15]

Because of the glut of applicants for Caltech and other positions, Bergmann had to settle, for the time being, for a less prestigious job. In 1941, he and his wife, Margot, moved to North Carolina and began working at Black Mountain College. An experimental college in a beautiful setting, Black Mountain proved a relaxing interlude where he could begin to establish his own identity, independent of his famous mentor. He worked in the physics department, while Margot taught chemistry. As their son Ernest related, "Black Mountain was a 'youthful experience' that they enjoyed recalling in later life."[16]

In the pastoral setting, Bergmann completed a masterful new textbook, *Introduction to the Theory of Relativity,* that proved vital for the dissemination of five-dimensional theory, among other subjects. Einstein kindly supplied the forward. Providing a basic but comprehensive overview of Einstein's work, "for a long time it was the book everyone read when they were studying general relativity,"[17] as physicist Steven Weinberg attested.

The book is divided into three parts: special relativity, general relativity, and unified field theories. Much of the third section describes Kaluza's approach and documents Einstein's attempts to extend his five-dimensional model. By placing higher-dimensional unification virtually on par with other topics in relativity, Bergmann publicized the importance of the subject. It was like an obscure documentary shown in tandem with a popular feature-length movie. Anyone drawn to the main attraction would necessarily be exposed to the more erudite material. For a full generation thereafter, practically every student of relativity would have at least some familiarity with Kaluza-Klein theory.

In 1942, Bergmann obtained a more research-oriented academic position at Lehigh University in Pennsylvania. Margot became a physical chemist at the Polytechnic University in Brooklyn, specializing in X-ray crystallography. To accommodate both of their careers they rented an apartment together in Manhattan. Savoring life in the Big Apple, they could spend time together on weekends, with each going his or her own way during the week. This became their commuting pattern for many years.

Teaching an undergraduate course in introductory physics, Bergmann was sometimes teased for his accent and his mannerisms. Chrysler CEO Lee Iacocca, perhaps Bergmann's most famous student at Lehigh, recalled the class in his autobiography:

> In my freshman year, I almost failed physics. We had a professor named Bergmann, a Viennese immigrant whose accent was so thick I could hardly understand him. . . . Somehow, in spite of my difficulties in the class, I got to be good friends with [him]. We would walk around the campus, and he would describe the latest developments in physics.
>
> There was something mysterious about Bergmann. Every Friday he'd end the class abruptly and leave until the following Monday. It wasn't until several years later that I finally learned his secret. Given the nature of his interests, I probably should have guessed. He used to spend every weekend in New York working on the Manhattan Project. In other words, when Bergmann wasn't teaching at Lehigh, he was working on the atomic bomb.[18]

Ernest Bergmann found this account somewhat far-fetched. "I read the autobiography and asked my father about it. My father was *not* Viennese. He did go into New York when he could because my mother was staying there expecting me (I was born November 1942). So, *I* was the 'Manhattan Project'. My father did not remember Iacocca."[19]

Another legendary tale about Bergmann's Lehigh days has at least some element of truth: "Lecturing to a class on the fourth floor, Bergmann sat gently rocking on the wide ledge next to an open window, when suddenly he rocked too far and disappeared from view. As the class sat stunned in horror, a hand reached in still holding the piece of chalk, Bergmann crawled back and calmly continued with the lecture. According to the story, workmen had erected a scaffold which provided an ample platform just outside the window."[20]

After leaving Lehigh in 1944, Bergmann joined the war effort as a sonar analyst. This was his first applied work, well outside the scope of relativity, yet he handled it deftly. Then, in 1947, he joined the faculty of Syracuse University, where he would remain for thirty-five years, establishing the leading department for gravitational studies in the country.

During the same period, Valya Bargmann had a much less circuitous path than his former colleague. When his assistantship with Einstein ended, he found work lecturing at Princeton University. He also did military research with John von Neumann related to gas dynamics. Then, after briefly working at the University of Pittsburgh, in 1948 Princeton appointed him to a tenured position.

Bargmann would teach at Princeton for many years, establishing a reputation as an adept scholar and teacher. His lectures on topics such as mathematical physics were "renowned for their clarity and polish."[21] The physicist Ken Ford, who attended his classes, recalled, "Bargmann was very prepared and well organized. He filled the blackboard from right to left with equations."[22]

Bargmann's wife, Sonja, became a valued editor and translator of the works of Einstein. She played a vital role in collecting Einstein's essays and other writings, reviewing their English translations with him and organizing them into a volume entitled *Ideas and Opinions.*

Pauli's Dreams

Pauli once suggested to Einstein that he thought Bargmann was more of a mathematician than a physicist. Pascual Jordan, he once remarked, was "only a formalist," implying that compared to a genuine physicist Jordan was a "lower form of life."[23] He told Oskar Klein that his true talent was for teaching, not for original physics research. And he sharply criticized Einstein for his vacillation. Who could possibly live up to Pauli's expectations?

As Stanley Deser recalled, "He was very rude and very self-assured. He was there for the greater glory of God. I have a letter from Pauli. I wanted to come and visit him at that time in Zurich. And he wrote me a reply saying unfortunately he doesn't issue visas from Switzerland so he can't stop me from coming."[24]

As a child prodigy who smuggled relativity articles into high school classes to read during boring moments, Pauli grew up with sky-high expectations for himself and others. These experiences led to a cocky, critical attitude, which his scientific peers generally tolerated because he was so brilliant. "I did talk to him quite a bit," Deser continued. "He was very, very smart."[25]

Pauli was especially close to his mother, and was devastated when in 1927 she committed suicide. Soon thereafter he got married, a disastrous experience that ended within a year in divorce. An ensuing drinking problem led him to seek psychotherapy.

At that time, the psychoanalytic movement was divided into two distinct schools of thought. While the mainstream, led by Sigmund Freud, delved into the role of the individual unconscious, a splinter group, led by Carl Jung, emphasized the power of the collective unconscious. In contrast to Einstein, who felt close to Freud, Pauli believed strongly in the Jungian model, hoping it would offer him interpretation and guidance in his times of need. From the early 1930s until he died in the late 1950s, Pauli corresponded with Jung and underwent psychoanalysis by one of Jung's assistants.

As part of the process, Pauli recorded more than a thousand of his dreams and sent off descriptions to Jung. Many of them had strange numerological elements associated with the numbers three and four. Jung analyzed and published some of these, using them to support his theories.

Pauli was discreet at first about mentioning his interest in Jungian psychology. Only late in his career did he publicize the connection, announcing his views that the science of the future would incorporate aspects of both physical and psychological phenomena.

Jung's archetype of the self, drawn from Eastern philosophy, is a union of opposites. In his statements and actions, Pauli exemplified this principle. Although he didn't change directions quite as much as Einstein, Pauli baffled his colleagues with his own share of contradictions.

Pauli's attitude toward unified field theories is one of the areas in which he wavered the most. On the face of it, he championed the view that the search for unification was futile. Mocking other physicists' quest for an all-embracing theory, he often said, "Men should not join what God has torn asunder."[26]

One of Pauli's first papers was a scathing critique of Weyl's theory. He similarly derided Eddington's unified approach. Though curious about Klein's original five-dimensional work, he encouraged him to give it up. And he branded Einstein's distant parallelism approach "terrible rubbish." No wonder Ehrenfest referred to Pauli as "the scourge of the Lord."[27]

Yet, in a strange turnabout, Pauli spent much of the mid-1930s working on his own unification proposal, a variation of Veblen and Hoffmann's projective relativity with some quantum aspects added in. He informed Klein about these ideas, encouraging the Swedish physicist's return to the subject as well. Pauli's interests also played a role, no doubt, in stimulating Einstein's studies with Bergmann and Bargmann. Pauli would later come to regret these detours, telling physicist Freeman Dyson, "If I had not wasted so much time trying to make sense of five-dimensional relativity (the Kaluza-Klein theory and similar attempts), I might have discovered quantum mechanics myself."[28]

Because he had Austrian citizenship, Jewish ancestry on his father's side, and fears about being expelled from neutral Switzerland to Nazi Germany, Pauli spent the war years safely in Princeton (aside from visits to the University of Michigan and Purdue). There he engaged in a brief but productive collaboration with Einstein. The results of their work had strong bearing on Einstein's decision to abandon his search for viable higher-dimensional models.

Einstein's Retreat

The research center where Einstein and Pauli would share much of the early 1940s was a much more verdant and spacious retreat than the old Fine Hall. In 1939, the Institute for Advanced Study finally moved to its own home, located just beyond the village of Princeton on the former Olden Farm. Encompassing many acres of wooded groves, fragrant gardens, and manicured fields, it provided the isolated, bucolic setting Einstein had missed from his days in Caputh.

Now his office, as well as his home, offered him the privacy that was essential to his peace of mind.

The main building of the institute, Fuld Hall, was tastefully constructed in the colonial style. While Fine was designed to be compact—with no office too far from the seminar room—Fuld enjoyed the luxury of copious space. Like an outstretched eagle, its wings spread out broadly from its central clock tower. In front of the building, a winding road placed further distance between the institute and Princeton University.

After all those years of correspondence (with the exception of a previous one-year visit by Pauli to the institute) it was a delight for Einstein and Pauli to be sharing the same roof. Despite his criticism of Einstein's later work, Pauli revered Einstein's youthful achievements and was a staunch defender of the importance of general relativity. He had begun to see some of the difficulties of reconciling general relativity with the quantum conception, giving him an appreciation for Einstein's concerns.[29] Moreover, in some ways he

Fuld Hall, the main building of the Institute for Advanced Study, once housed Einstein's office when he was searching for a unified theory of nature. More recently it has served as the research center for prominent string theorists such as Juan Maldacena, Edward Witten, Nathan Seiberg, and others.

had come to see himself as a second Einstein, and looked upon that period as a kind of changing of the guard. That time of closest friendship between the two was capped when Einstein nominated Pauli for the Nobel Prize, which he won in 1945. Pauli viewed a speech Einstein gave in his honor at that time as "like the abdication of a king, installing . . . his successor."[30]

Einstein and Pauli wrote a single paper together—a meeting of the minds to try one more time to implement Kaluza's notion. The paper acknowledges the beauty of the five-dimensional idea, while emphasizing the technical problems in getting it to succeed. "When one tries to find a unified theory of the gravitational and electromagnetic fields," they wrote, "he cannot help feeling that there is some truth in Kaluza's five-dimensional theory. Yet its foundation is unsatisfactory."[31]

In an approach that in some ways foreshadows the modern brane world model, Einstein and Pauli place no restrictions on the size of the fifth dimension, treating it as an equal partner with the others. They don't even curl it up into a tube. Instead, they assume that all fields are extended, like strands of spaghetti, into the five-dimensional space-time. Seeing only the tips of this "spaghetti" we would believe that space-time has four dimensions, even though it would actually have five.

Einstein and Pauli tried to find nonsingular (without infinities), stationary solutions to this model that would correspond to the familiar particles. When these conditions were injected into the body of general relativistic formalism, the results weren't particularly healthy. The only solutions that existed had zero mass and zero charge, precluding the possibility of electrons or protons. It was some strange universe, but it certainly wasn't ours. They presented the paper as essentially a negative finding—a scientific dead end.

Believing he could go no further, Einstein never returned to the subject of five-dimensional theories. The extra dimension that he hoped would work like a charm instead became just another millstone. From that point on, his unification schemes would be framed in ordinary space and time. In an autobiographical essay, written several years after the paper with Pauli, Einstein mentioned this decision: "I gave up an open or concealed increase in the number of dimensions, an endeavor originally undertaken by Kaluza that, with its projective variant, even today has its adherents. We shall limit our-selves to the four-dimensional space."[32]

After the war, Pauli returned to Zurich to resume his former pro-fessorship at the ETH. Circumstance would soon introduce him to yet another unified model, this time developed by his friend Pascual Jordan.

Conscience and Cosmos

During the Nazi reign, the German scientific community was hardly monolithic in its views. Depending on one's conscience and one's ambition, one could take a number of paths, ranging from protest-ing the regime to embracing it. Naturally, the former option was the most dangerous, leading to possible arrest, expulsion, or death. In contrast, by choosing the latter path—taken by fanatical anti-Semites such as Johannes Stark and Philipp Lenard—one could count on the government's delight but sheer revulsion abroad. A third, "middle-of-the-road" possibility, taken by Kaluza and many other academi-cians, was to cloister oneself in research and teaching, avoiding politics whenever possible.

Then there was Heisenberg, who kept up close foreign ties until the war started, hoping that Hitler's blazing star would burn out quickly. These international connections resulted in attacks on his character by Stark and others. During the war, however, he contro-versially worked in the Nazi atomic program. Though he later claimed that he tried to stall these efforts, letters by Bohr recently made public have cast such assertions into doubt. Many of Heisen-berg's colleagues were angry with him after the war. John Wheeler recalls that when escorting Heisenberg around Princeton, a number of professors, including Bargmann, refused to shake his hand. Bargmann just turned his back to him, Wheeler vividly remembers.[33]

If Heisenberg's behavior was confusing, Jordan's was downright bizarre. Even before the Nazis arose to power, he published ultra-nationalist opinions in a far-right journal. So as not to alienate his more mainstream friends, he used the pseudonym Domeier. Then, in November 1933, as Hitler's legions marched through the streets of Berlin and other German cities, Jordan joined a unit of stormtroopers. The friend of Klein, Pauli, Born, Bohr, and many other Jewish or half-

Jewish scientists was now wearing a "brown uniform, jackboots and swastika armband."[34]

While lending his voice to the Nazi insanity, Jordan maintained a staid physics professorship at the University of Rostock. There he chose research topics in the fields of quantum theory and relativity, condemned by those (such as Stark) attempting to "purify" German physics. Unlike most of his compatriots, Jordan continued to mention Einstein's name and contributions in his work, and argued vehemently against extending nationalism to science.

This latter stance won him the continued friendship of Pauli, who would play a considerable role in his rehabilitation after the war. Rather than condemning him, Pauli merely teased him about his Nazi-era writings. "Herr Jordan how could you write such a thing?" he once asked, to which Jordan replied, "Herr Pauli, how could you read such a thing?"[35]

In 1945, with the Allies closing in on Berlin, Jordan picked that moment to introduce a new theory based on the work of an Englishman, Paul Dirac—all this while working in a rocket center, doing meteorological calculations for the Luftwaffe. Then, during the Allied occupation of Germany, Jordan relocated to Göttingen, where he would develop his theory further while temporarily housed in its Mathematical Institute.

Pascual Jordan, who worked on a variation of Kaluza-Klein theory around the time of World War II, made his early reputation in quantum theory. In the eyes of many physicists, his standing fell when he joined the Nazi party. After the war, he reestablished his credentials as an innovative thinker in physics.

Jordan's approach, known as the scalar-tensor theory or the Jordan-Brans-Dicke model (for two other physicists who later rediscovered it), modifies general relativity so that the strength of gravity diminishes with time. It embellishes upon Dirac's 1937 idea that the current discrepancy between the magnitudes of electromagnetic and gravitational forces is due to the gradual weakening of the latter. Working out the figures, Dirac calculated that if Newton's gravitational constant, "G," decreased in proportion to the age of the universe, it would well explain why two electrons experience much, much greater electric repulsion than they do gravitational attraction. With no empirical evidence to support his notion except for pure numerical coincidence, Dirac put his idea aside for many years. But Jordan found it intriguing and wanted to pursue it further, connecting it with Pauli's version of projective relativity.

In Jordan's embellishment, the gravitational constant becomes an ever-changing scalar field, like a variable representing the fluctuating temperatures on a weather map. This transforms general relativity into a more complicated set of equations. No longer are the Einstein tensor (describing the geometry) and the stress-energy tensor (detailing the properties of the matter and energy) directly proportional. Rather, they have a more complex relationship.

Jordan's original manuscript describing his ideas was accepted by a German journal that had ceased publication by the end of the war. Fortunately, Pauli managed to obtain a copy of the galleys and hand them over to Bergmann in 1946. Bergmann was astonished to discover that Jordan's theory was almost identical to a model he and Einstein had developed, but had rejected and never published. After getting over the shock, Bergmann was heartened to see renewed interest in this topic. "The fact that another worker in this field had proposed the same idea, and independently, is an indication of its inherent plausibility," he wrote.[36]

Bergmann pointed out to Jordan that his model could be more elegantly expressed as a five-dimensional theory, similar to Kaluza's, but with a scalar field replacing the constant in the final (fifth row, fifth column) position of the metric tensor. If one imagines the metric tensor as a five-by-five checkerboard containing geometric rules, the scalar field would occupy a perch in the lower right-hand corner square. The placement of this scalar field ensures that gravity shrinks down over time. Bergmann published these ideas in his own paper on the subject.

Meanwhile, yet another researcher, Yves Thiry, had independently developed a comparable five-dimensional approach. Thiry, a student of the well-known French mathematician André Lichnerowicz, focused on a simple way of deriving Kaluza's equations and ended up reproducing Jordan's results as well. Scalar-tensor theories seemed to be popping up everywhere.

Jordan found out about Thiry's research in a roundabout way. Lichnerowicz told Pauli, who conveyed the news to Jordan in a letter.[37] Rather than being discouraged by the competition, Jordan pressed ever ahead, trying to find applications for his theory to astrophysics and cosmology—including ideas for how stars are formed and why the universe expands.

Living in Göttingen during the postwar years presented Jordan with the unique opportunity to discuss his theory with Kaluza himself. Because there was a terrible housing shortage at the time, Kaluza kindly arranged for Jordan and his family to reside at the Mathematical Institute. It was cramped and uncomfortable, even without any furniture at first. They cooked their meals in converted sanitary facilities, and slept with thin curtains dividing them. For working quarters, Kaluza offered Jordan a small room on the ground floor. Despite these hardships, however, Jordan appreciated Kaluza's generosity and welcomed his valuable advice.[38]

In 1948, Jordan went from rags to riches when he obtained a position at the University of Hamburg. Instead of living in borrowed office space, his family found ample accommodations in a posh suburban home. Wallpapering over his Nazi past, he joined the respectable Christian Democratic Party, and later became a member of parliament. Encouraged by Pauli, Jordan continued his gravitational research, aided by graduate students such as Engelbert Schucking. He put Schucking hard at work trying to find solutions to his strange equations. As a sideline he gave talks at public functions about topics ranging from science and religion to psychology and the paranormal.[39] He also published a popular book on his five-dimensional theories, *Schwerkraft und Weltall* (Gravity and the universe), that received mixed reviews.[40]

The Doors of Perception

Jordan's excursions into the world of the psyche were not surprising, given the heightened popularity of the subject during the postwar years, as well as his close friendship with Pauli. He and Pauli shared an interest in telepathic experiences, hoping that a single law of nature could explain the instantaneous communication of minds as well as particles. Pauli, in turn, was motivated by his ever-closer collaboration with Jung, applying quantum ideas to parapsychology. Pauli would contribute to Jung's book on the subject, *Synchronicity: An Acausal Connecting Principle*.

This interest in a "sixth sense" was closely connected with the scientists' fascination with the idea of unseen dimensions. In their mutual correspondence during the late 1940s and early 1950s, Pauli and Jordan would often mention Jungian ideas along with their discussions of extra-dimensional theories. During the same period other scholars, such as Kaluza and his assistant Gerhard Lyra, asked similar questions about possible connections between these fields. These scientists, to varying degrees, wondered if there might be a way of perceiving dimensions beyond space and time. Like the blind flower girl in Chaplin's *City Lights* for whom a medical breakthrough brings restored sight, could we somehow learn to detect realms that now escape our senses? In our dreams, or perhaps in certain transcendental experiences, might our minds be free enough to accomplish such a feat?

The philosopher Hans Reichenbach addressed the visualization of higher-dimensional spaces by drawing an analogy with the sense of color. He imagined a world in which the spectrum of colors represents the fifth dimension. If two billiard balls collided in such a world, their colors would become more and more similar as they grew nearer and nearer. Balls next to each other would have exactly the same hue. Reichenbach uses color merely to illustrate his point. His implication is that the fifth dimension could lie in a sense that we are missing, or perhaps one that we have yet to realize.[41]

Lyra, who did his doctoral work with Kaluza and then remained at Göttingen as a professor, saw fit to mix extra-dimensional theories with topics related to the paranormal. Encouraged by Jordan as well as

Kaluza, his more mainstream research included Lyra geometry, a variation of scalar-tensor theories that nicely encompassed Weyl's gauge method.[42] Lyra's extracurricular interests ran much broader, however, including exploring psychic experiences and investigating unidentified flying objects (UFOs). He once gave a talk to the German UFO Society in Wiesbaden: "About the necessity of a fundamental revision of contemporary physics for the understanding of parapsychic and cosmic phenomena—bridges to telepathy and the fourth dimension."[43]

Such connections are reminiscent of those made by the Society for Psychical Research in the late nineteenth century when it attracted followers of Zöllner such as William Crookes. Indeed the society witnessed a great revival of interest in the 1940s and 1950s, thanks in part to the famous experiments by American parapsychologist Howard Banks Rhine. Working at Duke University, Rhine tested the "mind-reading" abilities of a number of subjects, applying statistical methods to see if the validity of their guesses exceeded what one would expect from pure chance. Rhine coined the phrase *extrasensory perception,* or ESP, to describe his findings.

Pauli and Jordan were very much interested in Rhine's experiments, linking them with Jung's theories of synchronicity (acausal connections between coincidental events). They liked to compare notes with each other about the latest developments in this controversial endeavor.

In 1952, Pauli came across an item that particularly struck his eye. He mentions it at the close of a letter to Jordan: "By the way I also heard that in England wild mathematical speculations (in many dimensions) about Rhine's experiment appear in the Journal of the Society for Psychical Research. Do you know something about it?"[44]

The reference is to a paper by the psychiatrist John R. Smythies, entitled "Minds and Higher Dimensions."[45] Smythies interpreted Rhine's findings and other claimed parapsychological experiences within the context of a seven-dimensional model of the universe. Four of the dimensions house physical matter, while the remaining three contain "mind stuff": the rudiments of consciousness. These two worlds are causally connected but have their own internal rules. Thus, according to this approach, extrasensory perception would involve contact in the mental realm between people physically isolated in the material space.

Smythies spoke extensively about his theory with the English philosophers H. H. Price and C. D. Broad, each interested in parapsychology. During a visit to Zurich, he conferred with Jung about the nature of the collective unconscious.[46] More recently, he held discussions with physicists Andrei Linde and Bernard Carr about the role of thought in the universe. Keeping up with the latest Kaluza-Klein approaches, he has extended his model to thirteen or more dimensions—encompassing string theory and brane worlds.[47]

Smythies is perhaps better known for his work with Humphry Osmond, also in the early 1950s, on the nature of schizophrenia. He and Osmond, who coined the term *psychedelic*, conducted experiments in England and Canada using the hallucinogen mescaline, a derivative of the mescal cactus. Their controversial theory involved studying mescaline-induced visions for insight into the features of the schizophrenic state, with the ultimate goal of finding a cure.

One keen supporter of their work was the British writer Aldous Huxley, author of *Brave New World* and other novels. After Osmond introduced Huxley to mescaline, the writer began to advocate using drugs as a key to unlock new portals of sensation. Huxley chronicled his experiences in the 1954 book *Doors of Perception*. Although he doesn't explicitly discuss higher dimensions in his book, he describes how hallucinogens aided him in breaking down the walls of sensory limitations. This account helped renew the public association between dimensions and otherworldly experiences, especially as the psychedelic movement took hold in the 1960s. A prime example of this juxtaposition was *The Twilight Zone*, an American television series created in the late 1950s.

The Key to the Tesseract House

"There is a fifth dimension beyond that which is known to man," an eerie voice explained while fantastic images filled the screen. "It is a dimension as vast as space and as timeless as infinity." Thus began the first episode of *The Twilight Zone*, developed and narrated by Rod Serling and still considered one of the finest science fiction series ever made. For audiences transfixed by its bizarre tales of alien invaders, psychic powers, and hellish situations, the equation was simple: extra dimensions were infested with demons. Stay away from such bizarre realms if at all possible, the show's message strongly urged. Indeed in one of the subsequent episodes, "Little Girl Lost," a curious girl falls through a trans-dimensional portal in her bedroom and, to her parents' horror, becomes trapped. Her fate exemplified the lesson that hyperspace contact is not child's play.

These themes came as no surprise to the aficionados of the stories upon which many of the episodes were based. From the 1930s until the 1950s, science fiction was in a golden age, signaled by a proliferation of pulp magazines such as *Weird Tales, Astounding Science-Fiction,* and *Amazing Stories.* The speculative stories contained in these pages were the imaginative successors of the epics of Wells, Verne, and others. They enriched some of the older ideas with more up-to-date scientific references.

One of the pioneers of the pulp genre was the American horror writer H. P. Lovecraft. Honing his craft in the age of quantum physics, relativity, and multidimensional mathematics, Lovecraft incorporated aspects of these developments into his plots. Scientific verity, he believed, would help lure readers into accepting the more incredible aspects of his stories, setting them up for maximum shock.

Many of Lovecraft's stories featured an ancient race of beings, the Great Old Ones, that were waiting for just the right moment to reinhabit Earth. Apparently these creatures were familiar with Riemann and Gauss. Alluding to the distorted shapes and strange configurations of the cities constructed by these beings, Lovecraft presented non-Euclidian geometry as the building code of alien architecture. He also suggested that the creatures arose from "some vague realm or dimension outside our material universe."[48]

Lovecraft often envisioned higher dimensions as the byways of nocturnal wanderings. In "Through the Gates of the Silver Key," published in 1934, the main character seeks "to escape from the tedium and limitations of waking reality in the beckoning vistas of dreams and fabled avenues of other dimensions." While wandering through endless dreamscapes, he comes across the Ancient Ones. They explain to him "how childish and limited is the notion of a tri-dimensional world, and what an infinity of directions there are besides the known directions of up-down, forward-backward, right-left." The beings proceed to teach him these relationships. "The cube and sphere, of three dimensions, are . . . cut from corresponding forms of four dimensions, which men know only through guesses and dreams; and these in turn are cut from forms of five dimensions, and so on up to the dizzy and reachless heights of archetypal infinity."[49]

In "Dreams of the Witch House," the existence of extra dimensions presents an even greater source of terror. An overworked college student, immersed in the complexities of modern physics as well as the legends of supernatural practices, finds lodging in an "old Witch-House." In that spooky setting, he begins to make connections between the two disciplines, gradually realizing that witches knew more about higher mathematics than even "Planck, Heisenberg, Einstein and de Sitter."[50] As his knowledge of Riemannian geometry becomes substantial enough to impress even his professor, he discovers the possibility of hyperspace connections between distant points in the universe. Sure enough, he finds such a gateway in the Witch-House, drawing in unimaginable horror from the world beyond.

Although Lovecraft fancied himself somewhat of an academic, his knowledge came through self-education, not university training. The writer A. J. Deutsch, on the other hand, belonged to the Harvard astronomy department, and based his work on his direct experience. It was not *teaching* at Harvard, though, but rather *commuting* there that inspired his classic 1950 tale, "A Subway Named Mobius."

The Boston subway system is notoriously complex, consisting of many intertwined branches. Deutsch envisioned it becoming so tangled up that it spontaneously transformed itself into a multidimensional Möbius strip. While part of the system remained in the real world, a segment became hidden in hyperspace tunnels. Trains would whirl by, heard but unseen, because they rolled along tracks through a higher dimension.

A story with a similar theme, "And He Built a Crooked House," by Robert Heinlein, imagines a Claude Bragdon–like architect constructing a house in the shape of an unfolded tesseract. When fully unwrapped, a tesseract consists of eight cubes, shaped in the form of a three-dimensional cross. Conveniently, the architect designates each cube to be a separate room: living room, dining room, and so on. He is proud of how functional and efficient his layout seems to be.

The trouble begins when the designer takes the new owners on a tour of the house. Unbeknownst to them, an earthquake has collapsed the stretched-out tesseract into a fully four-dimensional hypercube. The cozy abode has now turned into a funhouse, with the rooms intricately connected in unexpected ways. By ascending three flights of stairs, for instance, the occupants find themselves back on the ground floor. Moreover, in one of the rooms, they discover windows looking out into eerie, otherworldly vistas. Fortunately, they all manage to escape before another seismic event dislodges the house from our space altogether, propelling it even farther into another dimension.

Yet another example of a science fiction story involving higher dimensions (there are many, many more) is Madeleine L'Engle's popular children's book *A Wrinkle in Time*, published in 1962. In this exciting tale, children are led by three mysterious women in search of their father, a scientist who has strangely disappeared to another planet. In a journey reminiscent of A Square's ascension in *Flatland*, the women transport the children off Earth and out of space-time by means of a hyperspace shortcut. They explain to the young adventurers how the process works: "Well, the fifth dimension's a tesseract. You add that to the other four dimensions and you can travel through space without having to go the long way round."[51]

These and other speculative stories introduced generations of readers—albeit in a fanciful way—to the curious higher-dimensional conceptions of the scientific community. Ironically, as these ideas gained ground in fiction, they seemed to lose momentum in science itself. For much of the 1950s and 1960s, relatively little work was published regarding five-dimensional theories of unification. While L'Engle's fresh new volume brightened thousands of bookstore shelves, the grand vision of Kaluza and Klein temporarily lay molding in yellowed journal pages. Particle physics was in its prime, and most theorists had other issues with which to contend.

CHAPTER 10

Gauging the Weak and the Strong

From time immemorial, man has desired to comprehend the complexity of nature in terms of as few elementary concepts as possible. Among his quests—in Feynman's words—has been the one for "wheels within wheels"—the task of natural philosophy being to discover the innermost wheels if any such exist. A second quest has concerned itself with the fundamental forces which make the wheels go round and enmesh with one another. The greatness of . . . gauge field theories is that they reduce these two quests to just one.

—ABDUS SALAM, Nobel Prize lecture, 1979

Parallel Histories

Scientific theories, like young children, experience growth spurts, then periods of relative quiescence. During the times when they shoot up, their development is obvious. Nevertheless, to build the internal structures required for their future success, they also need the quiet intervals. For Kaluza-Klein to become truly vital, a number of fundamental changes in subatomic physics would have to happen first.

After Einstein abandoned Kaluza-Klein in 1943 and the scalar-tensor theory of Jordan, Bergmann, and Thiry won few converts, discussions of higher dimensions would temporarily slow down—to be well outpaced by magnificent advances in particle physics and field theory. A new generation of theorists would grasp the reins of physics, baffling old-timers such as Einstein and Bohr with their radical notions and techniques.

One day Einstein was quietly relaxing in his home when he learned that he had a visitor. It was John Wheeler, eager to tap his views on a revolutionary new approach to quantum mechanics. Well aware of Einstein's skepticism, Wheeler nevertheless held out hope that the novel perspective could convince him. Einstein welcomed his colleague in, and the two sat down in the upstairs study to talk.

Wheeler carefully explained the "sum over histories" method, advanced by his student Richard Feynman. To calculate the results of any interaction between particles, according to Feynman, one can consider all the ways the particles might interact, multiply each by a likelihood factor, then add them up. For example, if one wishes to predict the chance that two electrons, after a collision, veer away from each other at a right angle, one can tally up all the possible ways they could execute this feat. Feynman developed the idea of sketching simple diagrams to make this tabulation.

The creepy part about all this is that it pretends that everything that *might* happen *does* happen all at once. If it applied to human interactions, it would be the pessimist's nightmare. Imagine if someone about to go for a job interview with a potential boss could picture every possible way the meeting could pan out. Then suppose the end result of that interview were somehow a weighted sum of all these possibilities. "There are more than a thousand different things I could say or do that would get me thrown out of the office," one might glumly conclude. "And every one would have the same unhappy result."

As intriguing as the notion of "sum over histories" is, Einstein was not impressed. "I still can't believe that the good Lord plays dice," he told Wheeler. After pausing for a moment to let his statement sink in, the elderly scientist continued, "Maybe I have earned the right to make my mistakes."[1]

Vanquishing Infinities

In 1948, Feynman employed his new method, diagrams and all, to develop an elegant resolution of one of the most daunting conundrums of *quantum electrodynamics* (QED): the quantum theory of electromagnetic fields. The issue, which Pauli and his assistant Robert Oppenheimer had wrestled with in the late 1920s and early 1930s, concerned the so-called self-energy of an electron. Under certain circumstances an electron can interact with the fields it produces. It can emit a photon, for instance, and then immediately reabsorb it. Such extremely short-lived "virtual photons" are a quantum consequence of the Heisenberg uncertainty principle's built-it fuzziness. The closer one probes to the electron's center, the greater the influence of this virtual photon cloud. If one calculates the total energy of such self-interactions using the Dirac equation and other quantum techniques, one obtains the highly worrisome answer of "infinity." To borrow a phrase from science fiction, that simply does not compute. Pauli, Oppenheimer, and others realized that infinite self-energy presented a major dilemma for QED, one that could doom the entire enterprise.

Independently, three physicists proposed solutions to this problem around the same period. The first was by Japanese theorist Sin-Itiro Tomonaga, who had the bad luck of arriving at his answer during wartime. Consequently, it wasn't published until the other two solutions appeared. The second and most detailed approach was offered by Julian Schwinger. Finally, the third and arguably the most intuitive solution was put forth by Feynman.

Their methods involved a technique, called renormalization, that uses a clever canceling out of infinities to arrive at an accurate, finite quantity for the electron's self-energy. During each step of the calculation, the infinite terms are arranged such that they precisely negate one another. It is in some ways a mathematical trick, but miraculously it works and conforms well to experiment.

One can view renormalization as a matter of accounting. Imagine a business, Quantum Enterprises, that starts out with $100,000 in its till. Every day the company gains $1,000 but also has to pay $1,000 in expenses. This exact balance of earnings and losses continues indefinitely. With concerns about its long-term future, the company calls in two different accountants (for independent estimates) to calculate how much money it will have many years down the road.

The first accountant is not too bright. As a first step he decides to compute the gains. He begins to add up all the earnings for not just years ahead but centuries as well. As he keeps entering figures into his calculator, the total gets greater and greater. Eventually, after he has computed the total for many, many centuries, the calculator overflows. "Infinity," he writes down as his answer. Then he decides to subtract each day's losses from the total. "Infinity minus $1,000 is still infinity," he writes down again and again. When it comes time to make his report, he announces, "I'm pleased to tell you that your company has the potential for infinite profits, if you continue your current policies indefinitely."

The second accountant is much more clever. She groups the figures in such a way that indicates their balance. Clearly the $1,000 lost subtracted from the $1,000 gained each day yields $0 per diem. The initial $100,000, plus an endless series of zeroes, just makes $100,000. Presenting the company with this finite figure, she offers them a more realistic view of their long-term prospects. Such is the superiority of "renormalization."

Schwinger and Feynman each presented his results at a conference in the Pocono Mountains of Pennsylvania. Along with many others, Bohr was there, curious to hear what the younger generation had to say. Schwinger went first, talking for a full five hours while covering all of the available blackboards with equations. Systematically, he showed one could tuck each infinity of QED under the rug by redefining certain terms. The observed mass of an electron, for instance, could be construed as already incorporating its self-energy. He took great pains to convince the audience that his methods were self-consistent, relativistically correct, and complete. Bohr and most of the other attendees were indeed impressed, though the sheer tedium of the exercise left many a little cold.

Feynman was next to speak. An iconoclast with a safecracker's take on the world's black boxes, he relished finding unique ways to resolve enigmas. Instead of garguantuan equations, he started drawing doodles on the board. Electrons he represented as straight lines with arrows, photons were squiggles, and so on. Whenever particles interacted with one another, he made their lines intersect. This was his personal shorthand—a technique that henceforth became known as Feynman diagrams.

Bohr was aghast at this cartoon version of quantum theory. Like a stern teacher, he explained to Feynman that the uncertainty principle doesn't permit drawing straight line paths for particles. There must be a built-in fuzziness. Feynman countered that he knew that already but was using the diagrams only as a bookkeeping device. Feeling misunderstood by the recognized leader of the quantum world, he left the conference very depressed.[2]

Hans Bethe, who consoled Feynman at the time, noted, "Schwinger deepened the existing theory, while Feynman invented a completely novel technique which at the same time simplified . . . actual calculations."[3]

Soon thereafter, Oppenheimer received a copy of Tomonaga's version. Then Freeman Dyson proved that all three methods were equivalent. Though Schwinger's detailed solution won him great respect, a steady influx of eager students at Harvard, and a Nobel Prize (shared with the two others), Feynman's elegant graphical explanation was to become the standard technique.

The Trouble with Gravity

Once so much was known about quantum electrodynamics, gravity seemed the shrinking violet in the corner. Physicists wanted to get to know her as well, but couldn't figure out how to approach her. Consequently, only a few dared try.

Peter Bergmann had just published what was to be his final research article on five-dimensional theory. A voice inside him, long muted by the demands of being Einstein's assistant, urged a return to his original ambitions. As a graduate student in Prague, he had deeply wanted to unify quantum theory with general relativity. Now that he was head of a whole new program at Syracuse—at that time the only center for gravitational studies in the United States—he finally had the chance to realize his life's dream.

Old habits die hard, however, and he felt obliged to run his idea past Einstein first. Predictably, his mentor had no interest. Engelbert Schucking reports what transpired: "Bergmann asked Einstein if he was interested in working on quantum gravity. Einstein said, 'You are now on your own.' Einstein wished him well."[4]

Bergmann then initiated what would be the first post-QED attempt to quantize gravity. In the early days of quantum theory, Rosenfeld had tried but failed, running up against infinite self-energy. Bergmann hoped there would be some way of circumventing this problem. His inaugural work on the subject, published in 1949, began with a manifesto proclaiming his far-reaching goal:

> At the present time, two great theoretical structures in physics can lay claim to containing significant parts of the "truth" which to unearth must remain the principal aim of both the experimental and the theoretical physicist. One of these structures is modern quantum physics as applied to both mechanical and field theoretical problems; the other is the general theory of relativity, which in the author's opinion represents the least imperfect "classical" (i.e. non-quantized) field theory.

As Bergmann realized, one of the major problems with uniting quantum theory and general relativity is the difference in how these models are based. Quantum theory concerns the evolution of a wave function over time. This wave function contains probabilistic information about the positions and momenta of particles moving through space. For general relativity, on the other hand, the principal entity is the metric, which describes space and time itself. The metric does not evolve over time. Because it is four-dimensional, it is rather in some sense timeless; thus it could not be subject to quantum dynamics in the same way.

Therefore, Bergmann believed, a key first step would be the search for the canonical variables of gravitation. These would be the parameters, analogous to position and momentum, for which Heisenberg's uncertainty principle and other quantum rules would apply. Replacing the timeless metric, these dynamical quantities would offer an equivalent but more quantum-friendly version of general relativity. In his paper, he didn't find these, but he felt like he was off to a good running start. He shared his pleasure with his true best friend, his wife, by dedicating the article to her.

As Bergmann's colleague Joshua Goldberg reported, "Peter inscribed his first paper with his white marker: 'Dearest Margot, with great love. It may be self-serving, but this is the first work of which I am myself proud.' With these words he left behind the prejudices of his mentor."[5]

Bergmann stayed in close touch with Einstein, however, keeping him abreast of developments in the field. Occasionally Peter and his family would drive down to Princeton in their beaten-up black

Pontiac and pay a call to the white-shingled house on Mercer Street that he knew so well. Bergmann's son Ernest recalled one such visit from his boyhood: "I remember them walking in a wild garden, perhaps Einstein's backyard, and talking about incomprehensible subjects. They seemed wrapped up in their discussions and not mindful of others. Einstein was badly dressed. It puzzled me that my parents were concerned about my clothing and grooming and here was a case of someone that was not setting a good example."[6]

Later, Bergmann would ask his mentor for a recommendation letter for a grant from the National Science Foundation to fund studies of quantum gravity. Einstein was quite obliging, despite his overall skepticism toward the subject:

> The application of Dr. P.G. Bergmann concerns a problem of central significance for modern physics. All physicists are convinced of the high truth-value of the probabilistic quantum theory and of the general relativity theory. These two theories, however, are based on independent conceptual foundations, and their combination to a unified logical system has so far resisted all attempts in this direction. If the decision were mine to make I should grant the funds . . . even though in my opinion the probability to obtain the great goal seems rather small at this point.[7]

Resuscitating Relativity

Bryce Seligman DeWitt arrived at Harvard pretty wet behind the ears. Fresh out of the Navy, he landed in one of Schwinger's first classes—on the subject of electromagnetism. It was a demanding course, to say the least. Like Kaluza, Schwinger lectured to his class with absolutely no notes, each session picking up exactly where he had left off the day before. During the final exam the class "sat in stunned silence" as they grappled with the monstrous questions only Schwinger could devise.[8]

With considerable mathematical talents, DeWitt managed to impress his teacher. He hoped to work with Schwinger, the only problem being finding common interests. Unlike most of his contemporaries, DeWitt relished the topic of general relativity. General relativity just wasn't done at Harvard—or anywhere else in the United

States at that time (except for Syracuse after 1949). If he wanted to pursue the subject further, he would have to find a way of integrating it with more current topics such as quantum field theory. As DeWitt recalled: "I had taught myself general relativity from Bergmann's textbook and I thought it was a very pretty theory. I chose Julian Schwinger for my thesis advisor and he had developed all these schemes for quantum electrodynamics. I thought, what the heck, let's see if we can't do the same for 'quantum gravodynamics'—in my naïveté. So I went to him, asked if I could do it and he said okay."[9]

Meeting with DeWitt perhaps "a total of twenty minutes altogether," Schwinger doled out his best advice on the subject. When DeWitt ran into trouble with an ambitious proposal to incorporate electrons, photons, and gravitation into a single theory, Schwinger advised him to simplify matters by leaving out the electrons. But mainly DeWitt was on his own. "I think I got good recommendations from him later because I left him alone," DeWitt conjectured.[10]

Stanley Deser was a later student of Schwinger, taking his preliminary courses just about when DeWitt was completing his thesis. Unlike DeWitt, Deser started out with no interest whatsoever in gravitation. Harvard was the home of particle physics, nuclear physics, and quantum field theory. Why would a young researcher hoping to make his mark wish to do anything else?

Graduating from Harvard and beginning a postdoctoral fellowship at the Institute for Advanced Study did little to change Deser's impressions. Oppenheimer, who had become the institute's director, expressed nothing but disdain for Einstein's seemingly pointless struggles with unification. For him Einstein's later endeavors were nothing but an embarrassment. "Oppenheimer warned us all against having anything to do with Einstein, because we might be struck by his pernicious interest in general relativity," recalled Deser.[11]

Nevertheless, Deser took the time to go to one lecture by Einstein. At first, because of Oppenheimer's warnings and the prospects of sitting through a talk in German, he was reluctant to attend. "Why should I waste an hour listening to this nonsense?" he told a colleague. Agreeing with Deser that the seminar would likely be worthless, the colleague nevertheless warned him, "If you miss seeing Einstein, what will you tell your grandchildren?" Swayed by sentimentality, Deser reconsidered. "So I went and it was more of this nonsense. And it was in German. And there were maybe 10 people in the room."[12]

Fate plays curious tricks. Though initially put off by gravity and unification, forces would conspire to make Deser one of the major players in those areas. And Deser insists that this had little to do with falling in love with Oskar Klein's daughter, Elsbeth, who was at that time visiting Princeton. They would continue their courtship in Europe, after Deser obtained a second fellowship in Copenhagen.

Meanwhile, John Wheeler, working at the other great center of learning in Princeton—the university itself—became similarly faced with gravitational theory's abysmal image. After making his name in nuclear theory, he wanted to investigate the subject that so enticed his aging neighbor on Mercer Street. There were no courses on the topic, so Wheeler decided to start one. He felt that the best way to learn general relativity would be by teaching it. His course became highly popular, with Bergmann's book as the text. Soon he became one of the leading experts on the subject.

Wheeler sometimes felt that his colleagues merely humored his newfound interest, without seeing much of a point to it. "The attitude toward relativity," Wheeler recollected, "was that everybody has his delusions and why not that."[13]

Throughout the 1950s and 1960s, Bergmann's Syracuse and Wheeler's Princeton formed the nucleus of a revival of general relativity in the United States. After obtaining a position at the University of North Carolina, Bryce DeWitt and his wife, Cécile DeWitt-Morette, established another leading center in Chapel Hill. Alfred Schild, a student of Infeld, set up yet another at the University of Texas. They were joined in their pursuits by Deser (who ended up at Brandeis), Charles Misner (who studied with Wheeler and ended up at the University of Maryland), Richard Arnowitt (who worked at Northeastern, then Texas A&M), Joshua Goldberg and Ted Newmann (who each worked at Syracuse under Bergmann), and many others. For most of these theorists, the question of how to quantize gravity and incorporate it back into the mainstream of physics was a pivotal concern. Only then, many believed, would it be realistic to contemplate unification of all the forces (through Kaluza-Klein or another mechanism).

Autumn Song

While Bergmann and Wheeler were shaping the American scene, Pauli had become one of the recognized leaders in European general relativity. Pauli's interest in Jordan's theory had given him motivation to immerse himself in the field. Moreover, since winning the Nobel Prize, Pauli felt a debt to Einstein that he wanted to repay. Honoring Einstein's vision would be the best form of tribute.

The fiftieth anniversary of the founding of relativity was coming up, and Pauli wished to pay his respects. He agreed to organize a major conference in Bern to commemorate the event. The Swiss government, and the European scientific community in general, were solidly behind honoring one of their "native sons," particularly because he had been so ill treated during the Nazi period. Einstein thought the idea was worthy and lent his support.

Unfortunately, by that time Einstein was in poor health. He had known for some time that his days were limited. In 1948, physicians had informed him that he had a grapefruit-size aneurysm in his abdomen. Its walls seemed relatively solid, so it seemed less risky to leave it alone than to operate. Over the years, however, it grew bigger and threatened to burst. Then in 1954, he became ill with anemia. In advance, he decided to send his regrets to the Bern committee, believing the perils of travel would prove too strenuous.[14]

As Einstein neared the end, another founder of unification theory would meet an untimely death. Theodor Kaluza left this world on January 19, 1954. Just about to retire, he suffered a massive heart attack while riding a bus home. Kaluza's assistant, Gerhard Lyra, was with him when he died. Kaluza had just recovered from a bout with the flu, and had decided to return to work. He and Lyra were administering an examination together, when Lyra noticed that the professor appeared tired. Lyra offered to take over the second part of the examination and escort him home on the bus.

As he sat down in his seat, Kaluza seemed unusually quiet. After a few minutes of silence, Lyra observed him close his eyes, roll his head back, and groan. Calling out to him and getting no reply, Lyra immediately asked the driver to stop the bus. The driver phoned a taxi to escort Kaluza to the hospital. Lyra held out hope until the end that Kaluza had just passed out and could be revived, but the physician who met him in the cab pronounced him dead. Apparently the flu had severely weakened his heart.[15]

History does not record whether Einstein had learned of Kaluza's death. At the time, it had been more than two decades since they had last corresponded. The inhuman treatment of Einstein and his colleagues during the Nazi period had dissuaded him from maintaining contact with many Germans. He stayed in touch with Max von Laue and several former associates, but few others.

Moreover, Einstein had long set aside Kaluza's model of unification in favor of other approaches. During his final years, he worked on what he called the "generalized theory of gravitation," a model that considered the metric tensor to be a combination of symmetric (rows and columns interchangeable) and antisymmetric parts. Associating the symmetric part with gravitation and the antisymmetric part with electromagnetism, he hoped to achieve unity within four-dimensional space-time. Not only was his theory anachronistic in content—completely ignoring nuclear forces, spin, and many other elements of modern physics—it resembled one of his 1920s models that failed. His final assistant, Bruria Kaufman, helped him with these calculations.

On April 13, 1955, Einstein's aneurysm ruptured. Although in great pain, he stoically refused surgery. "I want to go when *I* want," he told his doctors. "I have done my share, it is time to go."[16]

Over the following days his condition took varying turns. At one point, very close to the end, he was alert enough to ask for his writing papers and his glasses. As long as his hands could still move and his eyes could still see, he wanted to work on his unification theories. He died in the middle of the night on April 18.

Einstein never completed his dream of describing gravitation and electromagnetism by means of a single deterministic set of equations. Nevertheless, by lending his high public profile to discussions of unification, including the possibility of extra dimensions beyond space and time, Einstein brought greater awareness to the subject and helped launch the modern search for such methods. His vision has inspired many others to attempt unified theories of all natural phenomena, leading to today's multidimensional approaches, such as superstring and brane models. Einstein's notion of a Theory of Everything, elegantly explaining all aspects of the diversity and unity of nature, remains one of the most compelling goals of science.

Jubilee

When Pauli convened the "Jubilee of Relativity Theory" conference on July 11, 1955, it was a bittersweet experience for many of the participants. For those who knew Einstein well, including Pauli, Born, Weyl, von Laue, and the many former assistants who attended such as Bergmann, Bargmann, Infeld, Rosen, and Kaufman, his loss was excruciating. It was hard for them to think about relativity without recalling the warm, guiding presence of its brilliant author. Nevertheless, it was heartening for them to realize—by the sheer number of scholars represented as well as the broad scope of their work—that Einstein's spirit lived on in his disciples.

For some of the emigrant scientists, such as Bergmann, the jubilee represented their first trip back to Europe since the war. Bergmann brought his whole family and stayed for three months. He found it an opportunity to look up long-lost relatives in Germany—those who were lucky enough to have survived the Holocaust.

In the case of some of the other participants, such as Jordan and Klein, the conference offered a forum for discussing some of their more offbeat theories concerning unification and cosmology. Jordan talked about the changing gravitational constant, while Klein made two separate presentations. The first was related to the quantization of gravity, while the second concerned the behavior of galaxies in cosmology. As an alternative to Gamow's "Big Bang" model of the universe, proposed in the late 1940s, Klein had developed a theory of meta-galaxies: a finite, expanding collection of systems akin to the Milky Way. Following Klein's second presentation, British scientists Fred Hoyle and Hermann Bondi offered their latest thoughts on the steady state model of the universe, a more famous rebuttal to the Big Bang.

Yet other participants were from Paris, which had become a formidable center of relativistic studies. It would later assume a central role in the further development of Kaluza-Klein theories, particularly during the advent of eleven-dimensional supergravity in the late 1970s. At the time of the Bern conference, its leading lights were André Lichnerowicz and Marie Antoinette Tonnelat, each of whom wrote influential books on general relativity and the unification question. Each attended the jubilee and offered insights into their latest relativistic calculations.

Not all the attendees were well-established physicists. Some, such as Deser, were just beginning their careers, hoisting their sails in the winds of research, eager to head in promising new directions. Taking a break from his fellowship in Copenhagen, he was anxious to find out what was happening elsewhere.

"I was more or less really a tourist," Deser recalled. "I was driving around Europe and had noted there was this meeting coming up. I thought, of course, gravitation is some garbage that I had been warned against. On the other hand this was probably my only chance of a lifetime to get to see all these people."[17]

He was surprised to discover that the conference was in a natural history museum, where he had to push past "cases full of stuffed primates"[18] to enter the auditorium. "It was the weirdest meeting which I've ever attended. Everyone in the hall seemed to me to be 90, and I think some of them actually were, but most of them of course were not."[19]

At the conference Deser greatly enjoyed meeting Klein, his future father-in-law (who was then in his early sixties, one of the "young ones" by Deser's reckoning). They established a great rapport, and would consult with each other in future years about particle physics, gravitation, and other topics.

The conference closed with recollections of Einstein and expressions of mutual commitment to continue the international exchange of general relativistic ideas. Pauli expressed hope that the farewell to Einstein would mark a "turning point" in the history of gravitational theory. Sadly, Pauli died only three years after he made these remarks.

Getting Some Respect

Pauli was right on the mark about the relativistic revival. Slowly but surely, attitudes were changing. Some of the mainstream quantum physicists, such as Dirac and Feynman, started to appear at gravitation conferences and present insightful papers. Younger researchers were soon to follow. This heightened interest among bona fide quantum field theorists did wonders for the general perception of the field.

At the Stevens Institute of Technology, in Hoboken, New Jersey, the physicist Jim Anderson established a working group that would meet after the annual American Physical Society meetings. There, researchers such as Bergmann, Wheeler, Schucking, and DeWitt would chat and compare notes.

United, they combatted obstacles that each could not address alone. For example, they pushed hard for the right to have their articles appear in the major American physics journals, *Physical Review* and *Physical Review Letters*. In the late 1950s, the editor of these journals, Sam Goudsmit, made it clear that he considered articles grounded in experimental evidence, such as particle scattering measurements, nuclear decay models, and so forth, to be far more important than what he called fundamental theory. He published an editorial warning that effectively prohibited general relativity papers from the *Letters* journal.[20] Members of the Stevens group were duly concerned that their careful calculations would never appear in print in either publication. "One of us got wind of this," recalls DeWitt, "and I presented this at that meeting, and I think it was Wheeler who behind the scenes told Goudsmit to shut up."[21]

In 1957, Cécile DeWitt-Morette and Bryce DeWitt organized what would be the first of many regular conferences dedicated to general relativity. The meeting, which took place in Chapel Hill, attracted mainly American theorists. It had a younger, fresher feel than the jubilee, focusing on novel ideas to resolve many of the outstanding questions in classical and quantum gravity.

At the conference, Wheeler presented his latest ideas on a proposal for how to construct elementary particles from pure geometry. This so-called geon approach echoed the notions of Clifford almost a century earlier. Few of the conference participants accepted Wheeler's arguments. Bergmann expressed his strong concerns that the geon theory wasn't rigorous enough. Feynman teased his former supervisor about the concept, nicknaming him Geon Wheeler. "Hi, Geon," Feynman would call out every chance he could.[22]

Deser gave a talk, partly the result of discussions with Klein, about the role general relativity could play in helping to mitigate the divergence (infinite terms) conundrums in quantum field theory. Such a practical use for general relativity had enticed him to abandon his skeptical attitude toward the subject.

The Chapel Hill conference and a meeting the following year in northern France, each sponsored by the newly formed International Committee on General Relativity and Gravitation, precipitated significant progress in the quantization of gravity. In 1957, Misner proposed a way of applying Feynman's "sum over histories" method to gravity. The next year Dirac published a new formulation of general relativity based on the Hamiltonian (function expressing the total energy) approach. Arnowitt, Deser, and Misner followed up with their highly regarded ADM formalism, an extraordinary reworking of the structure of general relativity. The ADM method fulfilled Bergmann's initial goal of finding the canonical variables for gravitation. By slicing up four-dimensional space-time into three-dimensional spatial hypersurfaces, with each slice governing the shape of the next, it permitted researchers the luxury of considering gravitation as it evolves over time. Instead of viewing the universe as an unalterable roll of film, it gave them the option of editing it frame by frame.

Once the ADM team discovered the perfect slicing up of the universe, other theorists, such as Wheeler and DeWitt, could develop quantum visions based on arrays of alternative geometries. Every possible three-dimensional slice could be weighed according to its probability, like cheese in a deli. Then the true quantum universe would be a sandwich of these possibilities. The American theorists would construct such a model throughout the mid-1960s, resulting in the famous Wheeler-DeWitt equation of quantum gravity.

An Exercise in Higher Dimensions

Abdus Salam once said that higher dimensional unification theories go in and out of fashion. What is de rigueur one year might be a faux pas the next. No one could predict in which milieu this style of doing physics would show up next.

Following this maxim, the next development in the Kaluza-Klein approach would happen in an unexpected way: as a homework exercise in a French summer school. Granted, it was a school founded by Cécile DeWitt-Morette to advance innovative views in physics.

DeWitt-Morette established the Les Houches summer school, beautifully situated in the Alpine region of southern France, as a way of acquainting physicists with the most recent developments in the field. Each summer, the courses are based around a theme of current interest and taught by one of its leading practitioners. The school would help serve as the spawning ground for the next generation of theorists, including those who would revive the Kaluza-Klein approach.

In 1963, the central topic was "Relativity, Groups and Topology." Bryce DeWitt, teaching one of the classes, cleverly asked students if they could think of a way of combining Kaluza-Klein theory with Yang-Mills gauge theory. DeWitt had learned of Kaluza-Klein by reading Bergmann's textbook cover to cover. Yang-Mills gauge theory was developed by Chen Ning (Frank) Yang and Robert Mills as a possible model of the strong force. Based on $SU(2)$, a special group of two-by-two matrices, it attempted to show how protons and neutrons could transform from one into the other by means of non-Abelian (order of operations making a difference) rotations in isospace. This corresponded to the exchange of a massless gauge particle that presumably could be identified with the meson, even though the meson has mass. DeWitt advised students that merging the two concepts could lead to a unified field theory in which geometry could "perhaps provide the foundation for all of physics."[23] What an ambitious assignment!

DeWitt's problem foreshadowed the next phase in unification. With modesty, he could not see what all the fuss was about. "It seemed obvious. It was trivial if you had read the presentation of Kaluza-Klein theory in Bergmann's textbook. You just find the group manifold for Yang-Mills."[24]

Deser felt differently. "I was one of the few people, more than DeWitt himself, who was impressed by DeWitt's beautiful embedding of Yang-Mills plus Einstein in a higher-dimensional space. I thought it was really a validation."[25]

DeWitt's exercise did not herald the discovery of full unification, however. One of the reasons was that the original Yang-Mills model didn't adequately explain the strong interaction. As physicists came to realize in the 1960s, protons, neutrons, and mesons are not fundamental particles. They are composed of quarks, interacting with each other by means of gluons. The gluons, not the mesons, are the true carriers of the strong force.

An Assortment of Colors and Flavors

The quark model was proposed in 1963 by Murray Gell-Mann and Georg Zweig, each working at Caltech, to address the deluge of new particles discovered in cosmic ray showers as well as in particle-smashing accelerators. Beginning in the late 1940s, physicists realized that the constituents of atoms were just the tip of the iceberg. The atomic trio of protons, neutrons, and electrons, supplemented by photons, muons, pions, and neutrinos (hypothesized in the 1930s but found in the 1950s), soon became joined by many others. Using a variety of detection devices, from photographic plates to "bubble chambers" of boiling liquid hydrogen, experimentalists encountered an incredible diversity of particles with various masses, charges, spins, lifetimes, and other properties. Most of these were unstable, seen only as fleeting energy peaks before they decayed into other particles.

With the spirit of botanists faced with a lush tropical garden, researchers began to classify the known particles into distinct categories. The broadest grouping has to do with obedience to the Pauli exclusion principle. According to this rule, particles with half-integer spin, called *fermions,* never associate with other fermions that have exactly the same quantum numbers as themselves. Like patrons dining at solitary tables in a restaurant, this forces other fermions to spread out. Thus fermions tend statistically to take up a lot of room in the space of all energy levels. Examples of fermions include protons, neutrons, electrons, neutrinos, and many other types of particles.

Particles with integer spins, called *bosons,* do not have to obey such a rule. They can cluster together in as large a group as they want. In a restaurant of such particles, they would be able to share huge tables, inviting all of their boson buddies to join them. Hence, in terms of energy levels, they tend to take up far less space. Such social beings include photons and a vast variety of mesons.

Fermions and bosons play different roles in the universe. While fermions form the building blocks of matter, bosons mediate the forces that cause substances to attract each other, repel each other, remain stable, or decay.

Another classification of particles regards their susceptability to the strong interaction. Some particles, called *hadrons,* tend to feel the strong force. Protons, neutrons, mesons, and many others fall into this category. Other particles, called *leptons,* are completely oblivious. These include electrons, neutrinos, and muons. Clearly, some fermions are hadrons, while others are leptons. Fermions that happen to be hadrons are known as *baryons.*

These categories also reflect varying masses. In general, baryons, meaning "heavy ones," tend to be the bulkiest. Leptons, meaning "light ones," comprise the lightweights. Mesons are midlings, generally weighing somewhere in between. Finally, photons possess zero mass.

Nestled within these classes include the so-called antiparticles. Based on a notion first proposed by Dirac, most particles have antiparticle companions. Though identical to their matched particles in mass and certain other properties, they possess several important differences. If the particle is charged, the antiparticle has opposite charge. If a particle and antiparticle collide, they annihilate each other, forming pure energy.

While detecting these different types of particles (and antiparticles) and noting their decay patterns, physicists discovered special conservation laws. They found physical quantities that didn't change, no matter what transpired. For example, total initial charge is always the same as total final charge. Other conserved quantities include baryon number, lepton number, and a property called strangeness. Gell-Mann proposed the strangeness number to explain why certain types of decay do not occur.

To organize the particle zoo, Gell-Mann and the Israeli physicist Yuval Nee'man independently proposed systems for arranging mesons and baryons into various multiplets. These multiplets are analogous to the isospin doublets that group protons and neutrons together in Yukawa's theory—extended by the added property of strangeness. The mesons fall naturally into one group, while the baryons separate nicely into several others. For example, the proton and neutron belong to a family of eight, including lambda, two types of xi, and three types of sigma particles. Each family shares certain traits including baryon number, spin, and approximate mass. Gell-Mann nicknamed this system the eightfold way after a Buddhist expression.

This scheme suggested a special symmetry that would permit transformations from one particle to another within a given multiplet. Just as Yang and Mills recognized that transformations between protons and neutrons could be represented by the $SU(2)$ group, Gell-Mann and Nee'man found that similar interfamily exchanges could be carried out by $SU(3)$. $SU(3)$ is a special group of three-by-three matrices. Although its simplest expression permits transformations between triplets of objects, in a different representation it can also act on larger multiplets.

The fact that $SU(3)$ can most simply be expressed in triplets led Gell-Mann and Zweig to yet another breakthrough. They discovered that baryons are each composed of three quarks or antiquarks. Mesons, in contrast, are each quark-antiquark pairs. This beautifully accounted for all of the family groupings in a much more elegant way.

Following the Caltech researchers' system, one can characterize each quark according to its "flavor." This has nothing to do with taste, of course, but is rather a fanciful way of making distinctions. They believed that three different flavors—called up, down, and strange—would be enough to classify all known hadrons. Since then, experimentalists have discovered many new hadrons, enlarging the set of known flavors by three more: charm, top, and bottom.

Another property of quarks is their "color." Baryons must have three differently colored quarks, and mesons, two. A separate $SU(3)$ group acts to transform one color into another by means of exchanging gluons. This tossing back and forth of gluons conveys the strong interaction. The theory of the strong force has consequently come to be called quantum chromodynamics (QCD), by analogy with the successful theory of electromagnetism, QED.

Breaking the Symmetry

The weak force once presented a similar puzzle. Physicists in the late 1950s and early 1960s had some sense of what a quantum field theory of the weak force might look like, but they couldn't get it to work. The simplest models, such as one proposed by Schwinger, just couldn't be renormalized. Then Sheldon Glashow, who had obtained his PhD under Schwinger, developed a novel way of combining the electromagnetic and weak interactions using a cross between the $U(1)$ and Yang-Mills gauge groups. Because electromagnetism is a gauge theory, it would only make sense for the weak force to be a gauge theory as well. This hybrid seemed quite promising, except for a fundamental dilemma. To preserve its mathematical symmetry, the Yang-Mills mechanism permits only massless gauge particles. Thus, if such an approach were to explain the weak force, physicists would need either to detect new massless particles, or come up with a way of giving the gauge particles mass without destroying the essential symmetry.

The first option was impossible; all the abundant massless particles could easily be detected and would be well known. Of the existing particles, only the photon and possibly the neutrino seemed to lack mass (we now know that it has a small mass). The photon was known to be the gauge particle of the electromagnetic force only. The neutrino couldn't be a gauge particle because it had all the wrong properties; for one thing it was a fermion, not a boson as required. Therefore no much massless carrier of the weak force could exist.

That left the second option: lending mass to the exchange boson. In 1964, Peter Higgs of the University of Edinburgh found a clever way of doing this: a method known as *spontaneous symmetry breaking*. This approach is based on the discovery that symmetries within one context might break down in another. A change in environment might precipitate a phase transition that converts a perfectly symmetric situation into one with a flaw. For example, a completely square block of pavement, cooled down suddenly during an icy January day, might spontaneously develop a noticeable crack. Its left side might then no longer be a mirror image of its right side, ruining its initial symmetry.

To establish the conditions for spontaneous symmetry breaking, Higgs added an extra field to the Yang-Mills equations. This Higgs field would fill the universe, reacting to changing temperature conditions. During the fiery early moments of the universe, the Higgs field would enjoy a state of perfect gauge freedom, able to assume any configuration in its internal space. Its gauge "pointer" could rotate in any direction, like a rapidly spinning roulette wheel. But then, as the universe cooled off, the Higgs field would undergo a phase transition to a different energy state. In doing so, it would lose its gauge degrees of freedom. Its pointer would be trapped in a single direction, like a roulette wheel that has run out of power.

As Einstein pointed out, although energy cannot be destroyed, it can transform itself into mass. The excess energy lost during the Higgs field's phase transition when its pointer gives up its ability to turn would convert into a certain quantity of mass. This mass, in turn, would become taken up by the gauge bosons that happen to be present. Thus, in short, the sacrifice of gauge freedom during the cooling down of space would lend mass to the exchange particles.

One might think of this scenario as akin to a water flume ride, with a circular boat coasting through calm waters before dropping down a steep hill. Imagine boarding the boat at the top of the hill, well before the plunge. You don't give much thought to where you sit. Because the boat is round, every seat seems perfectly equal. Indeed as the quiet waters gently spin the vessel, no direction appears better or worse than any other. This is analogous to the state of complete gauge symmetry.

But then there is a sudden "phase transition" as the craft begins its thrilling drop. Suddenly the water all seems to flow the same way, establishing preferential seating. At the boat reaches the bottom, the kinetic energy of descent converts into one giant splash. And guess who is in the front seat. As the relatively dry riders sitting in what has now become the back of the boat all laugh at your soaked clothes, you leave the boat feeling somewhat heavier. The breaking of the boat's circular symmetry has lent watery weight to your clothes. As in the case of the Higgs mechanism, a perfectly equal situation has spontaneously transformed into one with preferred direction, with the energy lost converted into mass gained.

The Higgs mechanism, combined with Glashow's proposal, provided ample stimulation for two bright young physicists: Steven Weinberg, then at Berkeley, and Abdus Salam of Imperial College, each working independently. In 1967, Weinberg and Salam combined the matter fields of weak-interacting particles with the Higgs field, the photon, and three new $SU(2)$ gauge fields—called the W^+, W^-, and Z^0—to design a successful unified theory of electromagnetic and weak interactions. After this marriage, the newlyweds assumed the shared appellation of the electroweak force. In a crowning achievement for science, two out of four of the natural forces were finally united as one.

Would this matrimony work, or would it run into problems? Many physicists adopted a "wait and see" attitude until its augers could be better read. Good fortune arrived with Dutch physicist Gerardus 't Hooft's comforting prognosis. Using an exciting new technique, in 1971 he found that spontaneously broken Yang-Mills theories were fully renormalizable. Any infinite terms that arose could be removed.

Soon thereafter, experimentalists working at the CERN accelerator near Geneva confirmed one of the Weinberg-Salam model's principal predictions. The Z^0 gauge field, essential to the theory, represented a new type of weak interaction that didn't affect the charges of the participating particles. At the time of the model's proposal, such a neutral current was purely hypothetical and had never been found. When the CERN researchers finally detected this chargeless exchange, their discovery appeared to validate the entire model. The eventual detection of the weak gauge particles themselves would complete the picture.

Successful in their matchmaking, Weinberg, Salam, and Glashow would soon apply their talents in attempts to craft further unions. In 1972, along with physicist Jogesh Pati, Salam was the first to try and unify the electroweak and strong interactions, within the context of a so-called Grand Unified Theory (GUT). Grouping quarks and leptons together, they hoped to find a gauge group that could convert one into another. Their predictions included the possibility of protons decaying into other particles, a hypothesis that is still being tested. Glashow, in collaboration with his student and future colleague Howard Georgi, made further attempts to find such grand unification. Weinberg has similarly explored such possibilities, speculating about the nature of a "final theory."

While all these matches were being made, gravity was temporarily left in the lurch. As 't Hooft, as well as Deser and his postdoctoral researcher Peter van Nieuwenhuizen, all demonstrated, ordinary quantum gravity could not be shaped into a renormalizable theory. Therefore, without major changes in how it was formulated, it could not join the other forces in a unified theory of nature just yet. Somehow the Clark Kent of gravity would have to acquire superpowers in order to combat the villains of irremovable divergences. In short order, the theory of supergravity would be popping out of a phone booth to the rescue.

Final Theories

In his later years in Stockholm, Klein maintained an active interest in particle physics. Consulting with Deser and others, he tried to keep abreast of the latest approaches, often making suggestions himself. Before Gell-Mann's schemes were perfected, for instance, he attempted to develop his own organizational principles for hadrons, somewhat resembling the SU(3) model that ultimately took hold. He proudly played an instrumental role in the awarding of a Nobel Prize to Yang and T. D. Lee for their discovery that the weak force violates a physical symmetry known as parity. This result became a vital aspect of electroweak unification. As new findings from particle accelerators streamed in, leading to ever-changing theoretical interpretations, Klein tried as much as possible to stay current in the field.

Another of Klein's pet subjects was attempting to find an alternative to the Big Bang theory. Finding the idea of a single moment of creation too unsettling, he tried to fashion a more conservative explanation. In 1962, in collaboration with Swedish plasma physicist Hannes Alfvén, he developed a model based on matter-antimatter annihilation. He persisted even after the discovery of microwave background radiation strongly supported the Big Bang approach.

As Klein grew older, his range of intellectual pursuits seemed to stretch out like the branches of a stately tree. He gave numerous lectures and authored papers in the fields of philosophy, religion, and the history of science. He was particularly interested in the lives and works of Galileo, Pascal, and a medieval scientist named Jordanus Nemorarius—about whom he published a well-regarded treatise. Hoping to help educate the public about scientific issues, he became a familiar voice on radio talk shows in Sweden.[26]

One of the few excursions Klein made in his later years was to the International Centre for Theoretical Physics (ICTP), established by Salam in Trieste, Italy. By the time of his visit, in 1968, the two scientists had become quite close. Despite vastly different backgrounds, they seemed to have a great deal in common.

Salam greatly admired Klein's five-dimensional theory. Unlike most of his colleagues at the time, he believed that higher-dimensional models, such as Klein's, stood a good chance of completing the unification program begun by gauge theorists. After developing one of his gauge models, he described it to Klein and wrote, "unless and until Prof. Klein produces a 5-dimensional or a higher time-space symmetry this is the final link in the chain."[27]

In addition to their mutual scientific interests, Klein and Salam shared strong humanitarian concerns. They saw it as their supreme duties to help the less fortunate. Klein played an exemplary role in helping fellow scientists resettle in Sweden during the Nazi period. Salam similarly aided impoverished Pakistani students in finding meaningful careers by establishing a scholarship fund at the ICTP. He also argued vehemently on behalf of improving the technological conditions of the Third World. Both Klein and Salam worked ceaselessly for world peace, seeking to find just resolutions to international conflicts.

The Six-Day War that took place in the Middle East in 1967 hit both of them hard. Its destruction seemed cruel and pointless, setting back efforts for scientific and technological advancement. With their own interfaith endeavors as models, they sought ways of convincing the Arabs and Israelis to set aside their differences and strive for common purpose. After discussing the matter in Trieste, Klein wrote to Salam about the possibility of further action: "During these weeks I have been thinking much about our talks in Trieste and especially of what you mentioned about a possible contact between Israeli and Egyptian physicists. Perhaps such an initiative would be helpful in the present situation."[28]

Until the early 1970s, Klein continued to be active. Then, declining health precluded him from engaging in all of his scientific, philosophical, and humanitarian passions. The will was there, but not the stamina.

Klein died February 5, 1977, in Stockholm. He led a full life and was proud of his achievements. Though his contributions to physics were vast and vital, it was often his personal warmth, keen intellectual curiosity, and heartfelt concern for others that left the most lasting impressions upon those who knew him.

Sadly, fate sometimes prevents prophets from entering their own promised land. If Klein had only lived a few years longer, he would have witnessed the modern revival of Kaluza-Klein theory. He also would have seen one of his humanitarian dreams come true—the peace treaty between Israel and Egypt. Finally, he could have been present at the awarding of the 1979 Nobel Prize in physics—the trophy for decades of unity attempts. He would have been delighted that the Nobel committee granted joint honors to Weinberg, Salam, and Glashow for their electroweak model.

Salam's entrance to the Nobel ceremony was quite sensational. Among corecipients and audience members wearing suits, he dressed to make a statement about non-Western values. As his former student Michigan physicist Michael Duff described it, "Salam arrived attired in traditional dress: bejewelled turban, baggy pants, scimitar, and those wonderful curly shoes that made him appear as though he had just stepped out of the pages of the *Arabian Nights*. The net result, of course, was that he completely upstaged Glashow and Weinberg."[29]

The content of Salam's Nobel lecture, although it began with historical allusions, ended on quite a revolutionary note. After describing the steps that led to the electroweak model, then detailing notions for grand unification, Salam addressed what he saw as the future of theoretical physics: the emergence of eleven-dimensional supergravity. Referring to work by the French physicists Eugene Cremmer, Bernard Julia, Joël Scherk, and others, he expressed his hope that their higher-dimension model could incorporate all known fields. Their rendition of the "Kaluza-Klein miracle," Salam suggested, could very well be the road to the realization of Einstein's greatest dream.

CHAPTER 11

Hyperspace Packages Tied Up in Strings

Steve Weinberg, returning from Texas
Brings dimensions galore to perplex us
But the extra ones all
Are rolled up in a ball
So tiny it never affects us.

—HOWARD GEORGI, particle theorist[1]

Mirror Worlds

Supersymmetry is one of the most audacious proposals in the history of modern scientific thought. Year after year since it was postulated, experiments have failed to demonstrate its existence. Accelerators have smashed countless particles, producing not a single supersymmetric companion in their debris. Yet many theorists find it so compelling they can scarcely believe the world could survive without it. They argue that today's accelerators are simply not powerful enough to do the job. No other physical theory has won so many supporters with so little experimental support, surviving instead on the basis of its own mathematical beauty and internal consistency.

Critics, such as Sheldon Glashow, have compared belief in supersymmetry to a kind of religious fervor. "It's just some kind of abstract elegance," he has said. "Many of my friends have been doing supersymmetry for the past twenty years. It's a vast endeavor. It's a fascinating theory. It's an ingenious theory. It has accomplished, in terms of explaining phenomena, absolutely nothing . . . the trouble is not one of these predicted particles has been found."[2]

Briefly, supersymmetry is a gauge relationship between fermions and bosons, similar to the isotopic spin symmetry that groups protons and neutrons. It postulates that every fermion has a bosonic companion, and vice versa. Special rotations in "superspace" impart half-doses of spin to fermions, liberating them like a tonic from their staid obedience to the Pauli principle. Tossing aside their preference to keep strictly separate quantum levels, they acquire the yen for congregating in packs. The converse effect turns sociable bosons into solitary fermions. Every particle Jekyll, it would seem, has his own particle Hyde, only a transformation away from showing his other face. For instance, fermionic electrons, spun about in superspace, become bosonic "selectrons," while bosonic photons become fermionic "photinos." (Adding the preface s- or the suffix -ino are the ways scientists have named the supersymmetric companions.)

Supersymmetry charged into the picture in the early 1970s to help rescue an imperiled model of the strong nuclear force, called hadronic string theory. Developed by the University of Chicago physicist Yoichiro Nambu and others, hadronic string theory attempted to describe the behavior of protons, neutrons, and other strongly interacting particles by use of massless, elastic connections. By replacing particles with one-dimensional bands, it offered a geometric means of understanding an even earlier approach, called dual resonance theory, put forth by the CERN physicist Gabriele Veneziano in the late 1960s. While Veneziano's model used a mathematical formula to make predictions about hadrons, string theory provided a more visual description based on the science of vibrations. It derived particle properties, such as mass and spin, from the various modes by which a string could oscillate.

One shortcoming of Nambu's hypothesis was that it only represented bosons. Bosonic strings constituted the connections to which strongly interacting fermions (quarks and antiquarks) were attached. The fermions hung like dumbbells on the ends of each bosonic strand. Nothing, however, modeled the fermions themselves.

Another issue involved the dimensionality of the theory. As Rutgers theorist Claud Lovelace demonstrated, bosonic strings were mathematically self-consistent only in twenty-six dimensions. Few could believe this wild-sounding result.

Precocious as a child, Lovelace had read the works of Einstein and Dirac when only sixteen and had tried to construct his own amateur unified field theories. Obtaining his degree under Salam, a subsequent research stay at CERN launched him into the world of strong interactions. Fascinated by the string model, he sought a means of eliminating strange, faster-than-light entities called "tachyonic cuts" that had poked their heads into the calculations. Lovelace found that the only way to ward off this conundrum would be to situate the strings in a twenty-six-dimensional manifold. In January 1971, he nonchalantly recorded in his notebook, "I think we need 24 spacelike & 2 'timelike' dimensions to get complete cut cancellation."[3] He tucked this finding into a small section of a publication.

Physicists leafing through the article and coming across such an enormous figure were truly staggered. "Lovelace's paper was quite a shock to everyone," recalls Caltech physicist John Schwarz, "since until then nobody considered allowing the dimension of spacetime

Claud Lovelace, discoverer of twenty-six-dimensional string theory, along with his pet bird.

to be anything but four. We were doing hadron physics, after all, and four was certainly the right answer."[4]

Lovelace reports that he "certainly didn't realize the significance of the discovery at the time." In fact, he thought it was quite silly. When describing the twenty-six-dimensional result to other physicists at conferences, they shared in his mirth. He recalls "getting loud laughter at a Princeton seminar when he mentioned it."[5]

Later in 1971, physicist Pierre Raymond of the University of Florida found a clever way of incorporating fermions into string theory. Bosons could become fermions by vibrating in a different way. This method suggested an essential symmetry between bosons and fermions, with one freely able to transform into the other. Thus, the necessity of extending string theory to include both kinds of hadrons—fermions as well as bosons—provided the impetus for supersymmetry. Moreover, as demonstrated by Schwarz and André Neveu, superstrings (as supersymmetric strings came to be called) required only ten dimensions, rather than twenty-six. Though most physicists of that era still found ten dimensions weird, they thought it a substantial improvement over the larger amount. Schwarz sought in vain for a viable string theory in four dimensions, finally accepting that ten was the minimum.

Hadronic string theory didn't survive long. Following the twin successes of the electroweak model—'t Hooft's renormalization and the discovery of neutral currents—the bulk of the physics community adopted the stance that the strong interaction could best be understood by means of a non-Abelian gauge theory in the style of Yang and Mills. It seemed only natural to take what worked for one set of forces, modify it as needed, and apply it to another. Consequently, quantum chromodynamics, with its $SU(3)$ kaleidoscope of color charges, became the standard way of analyzing the strong interaction. Hadronic string theory, and the dual resonance model on which it was based, soon became as obsolete as a 1905 Studebaker.

Yet, like the Cheshire cat's smile, the allure of supersymmetry persisted when all else faded. As in the ubiquitous yin-yang symbols that adorned 1970s fashion, its union of opposites seemed to fit in with the times. Theorists wondered if there might be some way of incorporating supersymmetry into more standard particle models—a goal achieved by Julian Wess and Bruno Zumino in 1973.

Thanks to these physicists' work, supersymmetry could be expressed in the language of ordinary quantum field theory: namely, the dynamical equations and Feynman diagrams for point-particle interactions. The adaptation was like immigrants slipping old-country slang into their new tongue. From that point on everyone—not just the string theorists—could speak supersymmetry.

Gravity, Like Magic

During the first two decades after the death of Einstein, few had dared to tackle the problem of uniting gravity with the other forces. Even at the Institute for Advanced Study, where the stalwart professor had ceaselessly tried to subdue the beast, hardly a sign remained of that epic struggle. Young researchers saw little reason to confront the behemoth of gravity when there were countless other denizens of the particle zoo much easier to conquer.

That was all to change with a momentous lecture given at Princeton by Schwarz in 1975. It was a bittersweet homecoming for him, having been denied tenure several years earlier by that department. As it turned out, the significance of his ideas for the future of theoretical physics would demonstrate how mistaken they were.

The talk unveiled a deep connection between supersymmetry and gravity that Schwarz had recently discovered along with French physicist Joël Scherk. Within the context of string theory they had tapped the secrets of curious massless bosons with twice the spin of photons. Because these novel bosons corresponded to no known particles, many string theorists had dismissed them as useless appendages. Scherk and Schwarz realized that these spin-two objects were not superfluous at all; rather they were the very carriers of gravity, called gravitons. Thus gravitation was embedded in the bosom of supersymmetry itself.

About the arty rock band the Velvet Underground, critics have remarked that few people bought their records, but every one of these listeners formed his own band. Similarly, Scherk and Schwarz's finding attracted little notice at first, but those who paid attention ended up shaping the future of theoretical physics. Among those listening carefully to their message were two eager young physicists,

Bernard Julia and Edward Witten, each of whom would help mold new higher-dimensional unified theories. They have each mentioned how the discovery influenced their careers, persuading them that general relativity was a consequence of a grander principle of nature.

Born in Paris on July 8, 1952, Julia grew up with a knack for calculations. Aspiring at first to be a mathematician, he decided to "do physics to be more useful more rapidly."[6] Developing a "fundamental interest in the way the world works," he set out on a course to apply his prodigious talents in group theory and differential geometry toward deciphering nature's hidden code. His studies took him to the École Normale Supérieure (ENS) and then to the University of Orsay. There he wrote papers on optics and on the electroweak theory.

The best part of Orsay, according to Julia, was meeting Neveu, Scherk, and Eugene Cremmer, pioneers of supersymmetry with whom Julia would eventually collaborate. A quiet, warm-hearted man with a hearty laugh, Cremmer shares Julia's Parisian background and fondness for music. (Cremmer enjoys listening to classical; Julia revels in belting out tunes on his clarinet.) Cremmer became interested in physics through reading popular science magazines as a child.

Scherk was similarly soft-spoken and friendly, but had a history of illness (related to diabetes) and depression.[7] Mathematics provided his main distraction from his many physical and psychological problems.

Through his contacts with Cremmer, Neveu, and Scherk, Julia realized that there was a world beyond particle physics, one with a symphony of vibrations that could possibly orchestrate all of nature's elements and more. Julia became entranced by the harmonies of string models and intrigued by the possibilities of supersymmetry.

Winning a grant to work at Princeton, Julia brought his "infection with supersymmetry" to the States. He soon became close friends with Witten, with whom he enjoyed sharing his newfound excitement. Handing Witten a review article by Scherk, Julia made the case that the unity of fermions and bosons would provide the keystone for more viable theories of the natural interactions.

A baby-boomer turned unifier like Julia, Witten was born on August 26, 1951. He grew up in the northwest suburbs of an ethnic, middle-class Baltimore wonderfully depicted by the filmmaker Barry

Levinson and others. There he acquired a progressive political out-look that he would continue to embrace as an adult. His father, Louis Witten, a gravitational physicist, was a recent graduate of Johns Hopkins and would become a professor at the University of Cincinnati.

Tall and lanky, with curly dark hair, thick glasses, a shy demeanor, and a soft falsetto voice, young Witten was the very model of the brainy pupil at school. His classroom performance was so extraordinary that he skipped several grades and was granted a gifted alternative curriculum. By high school, he excelled in so many subjects that he scarcely knew which to pursue. Not only was he fabulously talented in mathematics and physics, he also surpassed his peers in the humanities and social sciences. "We used to sit around when Edward wasn't there and talk about how he was the smartest person in the world," a former classmate recalled.[8]

When Witten began his undergraduate studies at Brandeis, he decided to major in history, with the aim of becoming a political journalist. He graduated in 1971, just in time to become involved in

Edward Witten, one of the leading theorists in superstring and M-theory. Witten follows in Einstein's footsteps at the Institute of Advanced Study in striving for a unified theory of all interactions in nature.

George McGovern's unsuccessful presidential campaign. Soon Witten realized that politics wasn't his true calling, at least as a career. He switched gears and decided to study physics at Princeton. Nevertheless, throughout his life he has kept up an active political agenda, focusing on efforts to resolve the Israeli-Palestinian conflict. In that manner, he has continued the tradition of Einstein, Klein, Salam, and other physicists who have strived for international peace.

When Witten arrived on the ivy-covered campus, he experienced quite a bit of culture shock. Princeton was far more traditional and remote than the liberal, metropolitan, ethnically diverse settings with which he was familiar. "Witten felt socially isolated at Princeton," recalled Julia. "Once I fed him a lot of cookies, when he was depressed."[9]

Around that time, Julia began to explore the curious possibilities of higher-dimensional models. Schwarz's talk motivated him to find out everything he could about extra dimensions, from whomever could answer his many questions.

On this topic Valya Bargmann, still at Princeton, proved an invaluable font of knowledge. Bargmann gave Julia references about Pauli's work in projective relativity, and relayed his own thoughts about unification. He counseled Julia about his and Pauli's later belief that higher-dimensional theories wouldn't work.[10]

Still, Julia wanted to learn more about the strange world of Kaluza-Klein. At a conference in Trieste, Julia would bluntly ask Dirac about the physical consequences of extra dimensions in quantum field theory. With obvious discomfort, Dirac indicated that it wasn't a good time to consider the subject. "Maybe later," he brusquely responded.

After obtaining a visiting research position at Harvard, Julia was invited to a formal lunch for the faculty. There he approached the field theorist Sidney Coleman and once again mentioned the fifth dimension of the Dirac equation. Looking at him aghast, Coleman responded, "I don't want to take any responsibility for this guy."[11]

Witten also moved to Harvard, working there for several years as a research fellow. He found the experience intellectually exciting, stoking the coals of his nascent interest in unification. Meeting a young Italian physicist, Chiara Nappi, who was also working there as a postdoctoral fellow, did wonders to ameliorate his loneliness. They would get married and move together back to Princeton. At the unusually young age of twenty-eight, he became a full professor of physics. She became a member of the Institute for Advanced Study. In his new position, Witten set off on a course toward becoming arguably the world's leading theorist, winning a steady stream of accolades for his extraordinary mathematical achievements.

Meanwhile Julia's search for fundamental truth had led him to become a leading researcher in one of the hottest new arenas of physics: the world of supergravity. Returning to the ENS in 1977, he lent his talent to helping explore this promising juxtaposition of supersymmetry, quantum theory, and general relativity.

The Great Race

The theory of supergravity had started off in a dash one year earlier. Once Wess and Zumino discovered that supersymmetry could be incorporated into ordinary quantum field theories, and Scherk and Schwarz pulled gravity out of supersymmetry's hat, it was only a matter of time before physicists raced to develop the first supersymmetric quantum field theory of gravity. Soon it was clear that two main teams were vying for the supergravity cup. One group, based at C. N. Yang's institute in Stony Brook, New York, was led by Peter van Nieuwenhuizen, Daniel Freedman, and Sergio Ferrara. The other team, housed at CERN, was headed by Deser and Zumino. Though there were many personal connections between the competitors— van Nieuwenhuizen was Deser's former postdoctoral fellow and Ferrara had worked closely with Zumino at CERN—sadly the race became bitter.

The two groups shared the same basic goals. Standard field theory techniques, of the sort perfected by Feynman and Schwinger, had proven inadequate for quantizing gravity. As shown time and time again—the latest being the 1974 paper by Deser and van Niewenhuizen—gravity could not be renormalized. "He and I had an enormous project trying to decide once and for all whether it would be possible to remove the infinities of general relativity if it were coupled to various types of matter. And of course there was no miracle," recounts Deser.[12]

Unlike the electromagnetic and weak interactions, horrific divergences seemed to be quantized gravitation's fatal illness. Nothing seemed to inoculate against them. Therefore, the quantum gravity program, as initiated by Bergmann, DeWitt, Dirac, and others, was on the rocks. Supergravity, with its natural definition of gravity from symmetry principles, seemed a promising new landscape to explore.

Just as broken symmetry of the combined $SU(2)$ and $U(1)$ groups brought the electroweak theory together into renormalizable form, the supergravity researchers hoped that broken supersymmetry would generate a renormalizable unified theory that included gravitation. They hoped that the shattering of supersymmetry would magically supply the correct masses for the known constituents of matter, while including massless messenger particles such as the photon and graviton.

"I like to think of supersymmetry as crutches," says Julia. "It is useful but you want to lose it. You need to break supersymmetry but not by brute force. You want to keep some of the nice features."[13]

Making use of the powerful transformational properties of the Higgs mechanism (bolstered to a "super-Higgs" mechanism), the teams aspired to generate familiar fields while keeping monstrous by-products—things never seen in nature—to a minimum. They hoped that the particle spectrum produced would well match experimental findings.

To implement their goals, the groups were faced with a challenge. Supersymmetry as then formulated was a global symmetry. This meant that it operated the same for all points in space-time, much like painting a canvas uniformly red. To create a gauge theory analogous to Yang-Mills, with symmetry breaking through a Higgs-like mechanism, they needed to make supersymmetry local. That is, they had to derive the particular transformational laws for each point in space-time. Like artist Georges Seurat, they needed to paint with dots and find the subtle positional changes in supersymmetry's manifestation. This would also bring supersymmetry into line with the local invariances of Einstein's theory of general relativity.

In spring 1976, after many months of excruciating calculations, both teams announced success and sent off their results to major journals. Immediately a priority battle erupted. The Stony Brook group felt that its ideas leaked out to the other group somehow. The Deser-Zumino collaboration, on the other hand, made a convincing case that its work was concurrent and independent.

Deser vividly recalls finishing a rewrite of their draft at four o'clock in the morning on a flight back to Boston from Europe. He was away from CERN for several weeks on a much-needed vacation. Upon his return, he left the paper on Zumino's desk to be sent out.[14] This slightly delayed the date in which the results were submitted, even though they were completed at the same time as the other team's. Most subsequent papers in supergravity have cited both group's achievements.

With the preliminary articles exuding the sweet scent of promise, researchers clamored to complete the project. There were myriad combinations of groups and frameworks to explore, any of which could conceivably unlock nature's vault of secrets. An exhaustive search began for the correct symmetry group and appropriate dimensionality that would render supergravity's unification mathematically vital and physically accurate.

The Eleven-Dimensional Creature
from the Left Bank

Parisian writers have always had a grim interest in what lurks on the less fashionable side of the Seine. The formerly impoverished Latin Quarter has been the setting for quite a few mysterious tales. Notre Dame's gruesome gargoyles no longer gaze across the river upon street urchins, cadgers, and other ruffled characters, but the labyrinthine streets in which they once walked still bathe in shadowy light. Though Hugo's hunchback is old news, a new legend has emerged about an eleven-dimensional creature invented in one of the labs.

The setting for this creation lies in the neighborhood of a domed classical structure called the Pantheon. One of the left bank's most splendid buildings, its dark catacombs provide the final resting place for some of the greatest French philosophers, writers, and scientists. Entombed in one of the subterranean alcoves is Joseph-Louis Lagrange, one of the founders of the fourth dimension. In life, his words inspired students to dream about another dimension beyond space. In death, his spirit has offered assistance to those wrestling with even greater dimensional realms. Reportedly, several theorists once jokingly visited Lagrange's grave to solicit his "advice" on a particularly difficult set of supergravity equations.

Just beyond the Pantheon is another maze of narrow streets. A left turn, then a right, and one encounters a modern multistory edifice, the Physical Laboratory of the ENS. Cremmer's office in that building is where in 1978 he, Julia, and Scherk first conjured up a supergravity that lives and breathes in eleven-dimensional space.

There were a number of reasons the French physicists decided to construct an eleven-dimensional model. Calculations by other researchers indicated that four-dimensional supergravity seemed difficult, if not impossible, to renormalize. Moreover, symmetry groups operating in conventional space-time appeared inadequate to encompass all the interactions of nature under a common umbrella. For full unification, a larger arena would offer considerably more opportunities.

Cremmer and Scherk had considerable experience dealing with ten-dimensional string models. In 1976 they had developed a pro-

The Physical Laboratory at the École Normale Supérieure. In this Parisian research center, Eugene Cremmer, Bernard Julia, and Joel Scherk developed eleven-dimensional supergravity.

cess called spontaneous compactification that employed a modification of the Higgs mechanism to explain why extra dimensions could not be observed. Their model involved a phase transition that would cause physical fields to lose their dependence on all but the four ordinary space-time dimensions. Instantly, equality between the dimensions would vanish, replaced by the segregation of the majority (six out of ten) into a compact space. Presumably, as in five-dimensional Kaluza-Klein theory, this compact region would be smaller than the Planck scale (the quantum theoretical length that serves as the lower bound for measurement), rendering it impossible to detect.

To utilize this mechanism, the French researchers attempted to construct a ten-dimensional version of supergravity that, after compactification, would break up into the gauge groups for the strong, electroweak, and gravitational interactions. But the ten-dimensional theory had a scalar field that created problems. Everyone became increasingly frustrated by attempts to resolve this issue. Finally Scherk asked the others, "What should we do to get this project finished?"[15]

The answer was to add one more dimension. This effectively cut out the scalar field, making the model more workable. As Cremmer pointed out, it would be relatively simple to reduce these eleven dimensions down to ten if theoretical reasons required it (for instance, to match up with string theory).

The group knew, based upon a recent result by German physicist Werner Nahm, that eleven dimensions would be the absolute upper limit for a credible "Theory of Everything." Beyond eleven, massless fields would appear that have spin greater than two. Since no such particles exist in nature, with the maximum being the spin-two graviton, one could rule out dimensions by the dozen or more.

Among the creators of four-dimensional supergravity, the eleven-dimensional version won almost immediate acceptance. "If I look back at myself," reflects Deser, "when I heard about 26 and 10 I sort of laughed a little, but not very hard, and remembered 5. But then when supergravity came and 11 were bandied about, that was perfectly normal."[16]

Schwarz, on the other hand, had reasons to think that eleven were too many. Joined in his quest by British physicist Michael Green, he held out hope that superstrings, not supergravity, would constitute the ultimate theory. "I must admit that when [eleven-dimensional supergravity] appeared I was a bit baffled," Schwarz remarked. "I felt that the only supergravity theories worth studying were those that might arise from string theories at low energy. However, since superstring theory only allowed ten dimensions, this did not seem to leave a role for 11-dimensional supergravity. I viewed it as a 10% error."[17]

The mathematically adept Scherk continued to be an asset to both the superstring and supergravity camps, serving as a vital link between them. But then tragedy struck. In 1980, Scherk, who had been plagued by illness and deeply depressed to the point of psychotic delusion, suddenly passed away at the age of thirty-five. He died from a drug overdose—very likely a suicide.[18] His own inner demons had proven far more formidible to him than the eleven-dimensional entity he helped create. For Schwarz, Cremmer, and Julia, who each worked with him day after day for many years, the loss could not be more painful.

Among the many tragic implications of his early demise, Scherk sadly missed out on discussions of the ramifications of his collaborative constructions. The mainstream physics journals soon became home to sophisticated analyses of how eleven-dimensional unification would work, including a seminal article by Freund and Rubin.

Shrink-Wrapped Spaces

Romanian-born theorist Peter Freund's interest in Kaluza-Klein theories began with a question he once asked during his freshman year of college. During a 1953 lecture by noted mathematician Gheorghe Vranceanu, Freund was intrigued by the statement that electromagnetism and gravitation could be united in five dimensions. Freund raised his hand and asked the speaker how the nuclear forces could be worked into the picture. When Vranceanu had no answer, Freund suggested increasing the number of dimensions to accommodate the extra interactions. In his zeal, he proposed that perhaps the dimensionality of the universe should be infinite.[19]

Freund's fascination with extra dimensions persisted through the many decades in which Kaluza-Klein theory was as fashionable as burlap clothing. Then when the bold hypothesis was on the brink of its comeback, it was time for him to make his mark. A 1975 article with Yong Min Cho showed how higher-dimensional, non-Abelian gauge theories could be compactified, expanding on work by DeWitt and others.

Five years later, along with his student Mark A. Rubin, Freund hit the jackpot. They conclusively proved that eleven-dimensional supergravity could compactify in only one of two possible ways: either seven dimensions shrinking down, leaving four large ones remaining, or four dimensions shrinking down, leaving seven large ones remaining. Presumably our own world is the former possibility, not the latter. They showed that other outcomes, such as only one or two dimensions collapsing, were completely out of the question.

Freund and Rubin's finding offered a splendid validation of the ENS group's brief history of the universe. According to this theoretical Genesis, in the beginning there were eleven equal dimensions. Living in this fleeting primordial paradise, all the known fields, fermions and bosons alike, interacted with each other in complete harmony, with absolutely no distinction among them. All possible interplays between these objects took place with equal likelihood and equal strength, leaving no one master or servant.

Then came the great Fall. The cosmos cooled down enough that it could no longer maintain its exalted state of sheer equality. It underwent a phase transition, assuming its true ground state, which turned out to be somewhat less than uniform. While the four dimensions of ordinary space-time remained expansive, the other seven dimensions curled up into a compact (closed and bounded) domain. Smaller than the Planck scale, these microscopic dimensions could no longer be observed from that point on. Supersymmetry collapsed as well, resulting in bosons and fermions assuming distinct roles. Over time, other gauge symmetries broke down too, eventually resulting in today's jumble of particles and wide range of interaction strengths.

Shortly before Freund and Rubin's article appeared, Yale physicists Alan Chodos and Steven Detweiler had proposed an entirely different explanation of how the extra dimensions might have shrunk down to minute proportions. Curiously their paper, "Where Has the Fifth Dimension Gone?" made no reference to eleven-dimensional supergravity at all. The theory was so new that they hadn't even heard of it. Instead it reached way back to the original Kaluza-Klein model, cleverly showing how it could be represented by a somewhat altered Big Bang model of the universe.

Chodos recalls how he and Detweiler put their minds together to create the first Kaluza-Klein cosmology since the days of Jordan. "What really got us going was that Detweiler, who was an expert in general relativity, knew about something called the Kasner solution to general relativity. It dates back to almost the same time as Kaluza's work. The Kasner solution is a very simple, time-dependent solution to Einstein's equations. It was never used for much of anything because it predicts that some dimensions are growing and other dimensions are shrinking with time. If you're only in four dimensions that doesn't make a whole lot of sense. But what we realized was that it fit rather neatly into a five-dimensional framework. So that's what got us interested in pursuing it."[20]

In other words, Chodos and Detweiler envisioned a rather strange Big Bang, one in which a quirk of geometry set the stage for the expansion of ordinary space as well as the contraction down to microscopic proportions of an extra dimension. The paper spawned a number of sequels tinkering with aspects of this scenario, including a supergravity-friendly rendition by Freund, as well as my own first research publication while a graduate student at Stony Brook. New to the field, I found Chodos and Detweiler's concept an intriguingly simple way to picture dimensional reduction.

A follow-up article by Chodos—this time with the Yale physicist Thomas Appelquist—involved yet another way of explaining why some dimensions are smaller than others. It relied on a quantum phenomenon known as the Casimir effect to induce the selective shrinkage. The Casimir effect involves an attractive force produced in a vacuum between two metal plates. The Yale researchers showed how the same effect would arise in a compact fifth dimension, causing it to contract down to the Planck length. According to Chodos, the paper attracted even more interest than his work with Detweiler. He was proud that it drew the attention of Weinberg and other veteran physicists. By that time, Weinberg, Salam, and some of the other pioneers of the standard particle model had become deeply involved in the pursuit of a viable Kaluza-Klein theory of unification.

The Chiral Calamity

In early 1981, Witten was about to turn thirty. At an age in which some theoretical physicists and mathematicians have already offered their greatest contributions (Riemann, for example), he was merely lighting the tinder for a brilliant career that has lasted decades and is still blazing with fiery intensity. Settled in Princeton with a tenured full professorship, he could enjoy the freedom to contemplate the curious geometries of multidimensional manifolds, searching for the one configuration that shimmers with the patterns of ultimate reality.

An article he wrote at that time, "Search for a Realistic Kaluza-Klein Theory," demonstrated his growing insight into the promises and pitfalls of the subject. It is a masterful exposition of the topic, grounded in history but branching out into aspects the originators of the idea never imagined. It begins with an extensive introduction, outlining for readers unfamiliar with Kaluza-Klein theory all the progress that had been made up until that time. Witten consulted carefully with Bargmann to make sure his chronology was complete.

After the historical remarks, the meat of the paper begins. Witten asks, "What is the minimum dimension of a manifold which can have $SU(3) \times SU(2) \times U(1)$ symmetry?"[21] In other words, how many dimensions are needed to accommodate quantum chromodynamics (the color-exchanging mechanism of quarks, gluons, and the strong interaction) as well as the Weinberg-Salam model (the field theory of the electroweak interaction). Only by including sufficient room for each of these symmetry groups, along with the space-time symmetries of gravity, could one construct an edifice housing all the natural interactions.

Witten's query is akin to the process a couple goes through when planning a wedding. The betrothed need to tally up how much space is needed for the bride's family and friends, the groom's family and friends, their neighbors, their parents' neighbors, their caterer, their caterer's neighbors, and so on. Only then can they establish the minimum size of the hall they need to rent.

Witten adds up the number of dimensions required for each of the gauge groups. $U(1)$, represented by motions around a circle, needs only one. $SU(2)$, epitomized by rotations along a spherical surface, involves two. Finally, $SU(3)$, exemplified by transformations in a space with several complex variables, requires four. Adding one, two, and four yields seven. Combining this with the four dimensions of space-time produces the value of eleven for the minimum number, precisely the dimensionality of supergravity. Taken in conjunction with Nahm's assertion that eleven is the *maximum* number of dimensions, this finding placed the theory of Cremmer, Julia, and Scherk on especially good footing. Their dice had rolled the lucky number.

For a couple getting married, booking a hall and realizing that it was indeed the right size comes as quite a relief. But then, well after the guests have left, arrive the more mundane worries having to do with establishing a new household. Similarly, after Witten resolved the number of dimensions issue, he raised a host of more technical dilemmas having to do with the chirality (right- or left-handedness) of fermions, the spontaneous breaking of all the different symmetry groups, the values of various constants, and so forth. Of these, he identified the chirality issue as the most formidable.

The chiral fermion dilemma concerns Yang and Lee's Nobel Prize–winning discovery that parity is violated during processes involving the weak interaction. Parity is a symmetry between right-handed and left-handed particles implying that if one exists, the other must be present as well. However, in one of nature's curious quirks, one class of fermions, the neutrinos, come only in left-handed varieties (that is, spinning in one direction relative to their motion); another group, the antineutrinos, exist only in right-handed form; while all the other fermions (quarks, electrons, muons, and the like) come in both varieties. The difference is akin to left-handed and right-handed screws that turn in opposite directions. When Glashow, Weinberg, and Salam developed the standard electroweak model, they built this distinction into the theory. Yet as Witten revealed, the eleven-dimensional supergravity model didn't take these differences into account. Parity was manifestly conserved, in glaring violation of known physical law. For supergravity to be all-encompassing it would have to be modified somehow to allow for chirality distinctions.

The problem is akin to an engineer designing a new divided highway system that permits driving on either side of the road. He builds ramps that lead from the right lane of one divided road to the left lane of another. It thus makes no preference for right- or left-sided driving. Critics immediately lambaste his scheme. They point out the highway code's "chirality restrictions" governing on which side of the road one can legally drive. They suggest that his "nonchiral" set-up could cause accidents. Belatedly realizing this, the engineer must modify his system by removing direct connections between lanes that ought to run in opposite directions.

The model proposed by Cremmer, Julia, and Scherk was elegant and simple. It seemed to have exactly the right number of dimensions. However, like a highway system that allowed every possible connection between lanes, it was just too symmetric. To match nature's inexplicable bias, it needed to be taught to distinguish between left and right.

Once Witten burst open the dam of high expectations, other major problems gushed forth. A 1983 paper he wrote with Luis Alvarez-Gaumé suggested that higher-dimensional supersymmetric theories suffered from a mathematical malaise called anomalies that simply could not be cancelled out. These blemishes violated physical principles such as conservation of energy, and allowed for strange negative probabilities. Imagine having *less* than a zero chance of being in a certain place.

For added worries, renormalization of supergravity appeared to be far more difficult than originally thought. Some of the gauge groups most compatible with the requirements of supersymmetry seemed the wrong sort to accommodate the standard model. Finally, a nagging problem with an unreasonably high cosmological constant (an antigravity term in general relativity introduced but then ultimately discarded by Einstein) found little hope for resolution.

Throughout the early 1980s, supergravity experts worked with ceaseless energy in attempts to resolve these difficult issues. In addition to the field's founders, numerous young researchers from all over the world joined in on the effort. British physicists dedicated to supergravity during that period included Michael J. Duff (a former student of Salam and postdoctoral assistant to Deser), Gary Gibbons, Peter West, and many others. Among the many contributors to the subject, other workers included François Englert from Belgium, Bernard de Wit from Holland, Hermann Nicolai from Germany,

Riccardo d'Auria and Pietro Fré from Italy, Bengt E. W. Nilsson from Sweden, and Christopher N. Pope, S. James Gates Jr., and Warren Siegel from the United States.

As a graduate student at Stony Brook during that period, I was astonished by the backbreaking level of endurance mustered up by the supergravity researchers. Vacations, holidays, evenings, weekdays, and weekends served equally well for excursions through endless vistas of Feynman diagrams. And van Nieuwenhuizen, as the group leader, seemed the hardest working of all. It was not uncommon to see him appear at the Institute for Theoretical Physics in the middle of the night, fresh off the plane from a conference but eager to attend to other duties. I remember that one time he returned from his native Holland sometime in the wee hours with only Dutch guilders in his pockets, not dollars. When I lent him a few dollars, he insisted on giving me guilders in exchange. The next day we traded back. Given the level of effort going on within those walls, I felt privileged in my own small way to have momentarily aided the cause.

Good Vibrations

For almost a decade, the glamour of supergravity lured some of the best graduate students, postdoctoral fellows, and young professors. Even as flaws emerged, these researchers clung to their cherished jewel, hoping that a bit of polish would make it shine once more.

During the same period, superstring theory was the poor, obscure cousin. Though it had once provided the motivation for supersymmetry, its underlying propositions seemed far too radical. Why exchange familiar particles for bizarre vibrating strings, field theorists felt, given the unequivocal success of the standard electroweak model?

The level of support for superstrings was so low that John Schwarz, one of its principal developers, clung to a nontenured position at Caltech for many years. Michael Green, its other main advocate, remained a little-known professor at Queen Mary College in England, where alternative theories were better tolerated than in the States.

In 1984, Green and Schwarz put their minds together and made one of the greatest breakthroughs of their careers. Through meticulous calculations they demonstrated that at least one version of ten-dimensional superstring theory was wholly free of anomalies. They had already proven that their scheme had no infinities and did not need to be renormalized. (The lack of infinities stemmed from the fact that strings were extended objects rather than infinitesimal points, and therefore did not require theorists to divide by zero in their calculations.) These were two major advantages over supergravity. When they made the announcement at a conference in Aspen, the audience was flabbergasted by their model's new look. Like a frumpy student dolled up for the prom, suddenly their theory attracted eager glances.

In Princeton, quickly replicating their feat, Witten declared his enthusiastic support. The hearty endorsement of a recognized mathematical genius soon won over many notable physicists to the world of strings. While Gell-Mann offered his blessings, calling it a "beautiful theory," Weinberg proclaimed that it was the "only game in town."[22] The same physicists, in contrast, viewed supergravity like yesterday's cold supper. "Eleven-Dimensional Supergravity (Ugh!)," said Gell-Mann at a conference.[23] The first superstring revolution had begun, with Schwarz, Green, and Witten leading the charge. (It is now called the *first* revolution to contrast it with the advent of M-theory, known as the *second* superstring revolution.)

Newcomers to the topic quickly acquainted themselves with its major elements. Instead of point particles, the fundamental building blocks of nature are vibrating strings. These come in two basic varieties: open and closed. Open strings resemble cut pieces of twine, each with two free ends. Closed strings are more akin to rubber bands, forming complete loops.

The size of each of these varieties is astronomically small (10^{-33}cm)—far more minute than anything previously known in nature. More than a billion, billion quadrillion strings would line up to form a single inch. This scale is so miniscule that if a hydrogen atom were blown up to the proportions of the Milky Way galaxy, the strings within it would only be the size of dust mites. Thus no known experiment could directly reveal the stringlike properties of particles.

The strings generate the disparate properties of nature much in the same manner that Jimi Hendrix produced his wildly diverse strains. Whenever a musician plucks a guitar string, it can vibrate only in fixed harmonic frequencies—exact multiples of a fundamental mode. The pattern of oscillations has either one peak, two peaks, three peaks, or, in general, any whole number. Because the ends are attached, fractional numbers of peaks are strictly forbidden. In music this creates distinct sounds; in physics it yields discrete energy levels. De Broglie applied this principle to electrons circling atoms in his successful quantum description. Now string theory suggests that electrons (and other particles) themselves constitute much, much smaller types of vibrations.

In the string model, each of the essential properties of elementary particles—mass, charge, spin, and the like—corresponds to various oscillations and configurations. The same string, vibrating in a different way, produces a completely different particle. For instance, the more massive muon comprises a higher-frequency oscillation than the electron. Bosons and fermions each pulsate in a characteristic manner. Because there are an infinite number of ways strings can oscillate, they can potentially produce any known particle in nature, as well as many never seen.

Much like tuning a guitar, altering a parameter called the string tension controls the range of possibilities for vibration. The stronger the tension, the harder it is for the string to oscillate. This creates more energetic modes of oscillation, reflected in more closely spaced wave patterns.

A major difference, however, between superstrings and guitar strings (aside, of course, from size) is that superstrings are free to move. In fact, they have many more directions of motion because they live in a higher-dimensional space. As recorded in an expanded space-time diagram, their movements trace out "world sheets": generalizations of particle paths. A world sheet appears like an undulating curtain or tube, woven strand by strand by strings evolving over time.

Strings can merge and divide in numerous ways. These fleeting liaisons are the equivalent of particle collisions and decay. In stringy Feynman diagrams these appear as world sheets joining together and then separating over time. String theorists use such tools to try

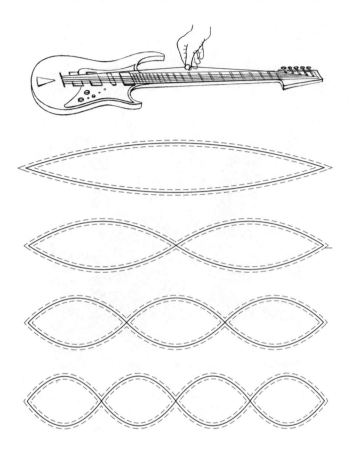

*Plucking a guitar string generates patterns of standing
waves, each with a whole number quantity of peaks.
Superstring theory, involving vibrating strings
far too minuscule to detect, associates each
pattern with distinct particle characteristics.*

to model how strings interact, hoping to match these portraits with
natural processes.

   The principal developers of superstring theory received numerous accolades for their achievement. Schwarz was finally given tenure
at Caltech, more than a dozen years after Princeton had denied him
the same privilege. He would eventually move into Feynman's office
after the esteemed theoretician passed away in 1988. In 1989, the
International Centre for Theoretical Physics awarded Green and

Abdus Salam, a codeveloper of electroweak gauge theory,
proudly watches John Schwarz, the cofounder of
superstring theory, as he displays his Dirac medal.

Schwarz its prestigious Dirac Medal. Four years later, Green would be honored with a full professorship at Cambridge. Witten received the Dirac Medal in 1985 and the Fields Medal, the mathematicians' equivalent to the Nobel Prize, in 1990, among his many honors. The MacArthur Foundation had already granted him its genius award in 1982. It would bestow the same prize to Schwarz in 1987.

An Embarrassment of Riches

Within a year after Green and Schwarz published their paper on anomaly-free superstrings, a number of other physicists discovered a host of additional viable models. Each was based on a different set of symmetries and had its own distinct properties. Soon researchers narrowed the scope down to five. These five variations are called the Type I, Type IIA, Type IIB, Heterotic-O, and Heterotic-E string theories. Type I, the original model, includes open as well as closed strings. The others permit only closed strings. Type IIA is nonchiral; all the others allow for handedness distinctions.

Working in Witten's department, the four discoverers of heterotic string theories—David Gross, Jeffrey Harvey, Emil Martinec, and Ryan Rohm—have been aptly named the Princeton string quartet. In concert, they found an ingenious way of blending separate string theories to form a greater harmony.

The term *heterotic* is an old-fashioned description from biology. It signifies the tendency of hybrid offspring to have qualities superior to their parents. In the case of strings, the appellation represents the combination of a supersymmetric theory with a purely bosonic theory to form a whole greater than the sum of its parts.

For a closed string, waves can travel either clockwise or counterclockwise around the loop. The two possible directions make it a chiral theory, suitable for modeling nature's disparity between left- and right-handedness. This is like a country dance with two concentric rings: the men circling in one direction and the women in the other. Replace the men with supersymmetric strings living in ten-dimensions, and the women with bosonic strings living in twenty-six-dimensions, and one has a good picture of the hootenanny orchestrated by the Princeton string quartet.

In order for the "dance partners" to be well matched, the bosonic strings must hide sixteen of their twenty-six dimensions. These extra dimensions must curl up into a compact space. Though the extra space is miniscule, it possesses ample types of symmetry. Remarkably, as the Princeton researchers found, it contains enough gauge fields (each corresponding to a particular symmetry) to reproduce the standard model of particles—with many more to spare.

Meanwhile Witten, along with Philip Candelas of the University of Texas at Austin and Gary Horowitz and Andrew Strominger of the University of California at Santa Barbara, discovered the proper class of geometries for the curled-up dimensions of string theory. Unlike Oskar Klein's model, which envisioned the extra dimension as a circle, and higher-dimensional generalizations of Kaluza-Klein theory that pictured squashed up hyperspheres and hyper-doughnuts, superstring theory called for a more unusual configuration. The correct way to compactify the extra six dimensions of ten-dimensional superstring theory, the researchers found, is into pretzellike shapes called *Calabi-Yau spaces*. Named after University of Pennsylvania mathematician Eugenio Calabi and Harvard mathematician Shing-Tung Yau, tens of thousands of varieties of such shapes exist. Each is twisted into its own distinct form, like the gnarly Joshua trees of California. The specific topology of a given Calabi-Yau shape— especially its number of holes—govern the symmetries and other properties of the physical model it represents.

With five viable superstring theories, each containing hundreds of gauge fields and tens of thousands of ways these models could be compactified, superstring theorists reveled in a bounty of mathematical riches. They hoped that soon scientific constraints would whittle these choices down to the one Theory of Everything.

When, by the late 1980s, none of the models could be eliminated, the wealth of possibilities became a major embarrassment. Nature needed a single theory to describe it, not a plethora of indistinguishable options. Moreover, no experimental results supported any of the theories—not even supersymmetry in general. Many predicted particles simply couldn't be detected. Skeptics such as Glashow stepped up their criticisms, urging the theoretical community to refrain from placing all its eggs in one Calabi-Yau-shaped basket. At a conference on unification he read a poem that ended with these lines:

> The Theory of Everything, if you dare to be bold,
> Might be something more than a string orbifold.
> While some of your leaders have got old and sclerotic,
> Not to be trusted alone with things heterotic,
> Please heed our advice that you too are not smitten—
> The Book is not finished, the last word is not Witten.[24]

Feynman, in one of his final interviews, expressed the viewpoint that superstring theory was nonsense. "I have noticed when I was younger that a lot of old men in the field couldn't understand new ideas very well . . . such as Einstein not being able to take quantum mechanics," said Feynman. "I'm an old man now, and these are new ideas, and they look crazy to me, and they look like they're on the wrong track."[25]

It was around that time that loop theory, largely developed at Syracuse, emerged as an alternative quantum theory of gravity. It did not require a belief in vibrating strings, supersymmetric particles, or squashed extra dimensions, rather just a fresh way of looking at general relativity itself.

Beyond Geometry

For the many decades when he headed the Syracuse relativity group, Peter Bergmann commuted each week from the Upper West Side neighborhood of Manhattan, first by train, then later by plane. Especially by rail, the trip was a long one, stopping first at all the major Hudson River towns, then all the locales along the Mohawk. By the time the train pulled into Syracuse station, much of the day had passed. While some might have bawked at such a drawn-out commute, Bergmann enjoyed having ample time to read and think. Moreover, in either direction the destination made it well worth it, be it his talented colleagues and students on the one end, or the radiant faces of his beloved Margot and their two boys on the other.

The Einstein centenary celebration in 1979 and his official retirement in 1982 were cause for reflection on Bergmann's own life's voyage. Appearing at all the major commemorations of Einstein, he reminisced about their joint work in five-dimensional unified field theory. He spoke about the subsequent progress that had been made in quantum gravity and speculated about future directions for the field.

Although Bergmann and the Syracuse group were curious about supergravity (and superstrings), even writing some papers on the topic, they were disinclined to jump on its speeding wagon. Their approach emphasized instead the step-by-step progression of developments in quantum gravity, each based on new insights into Einstein's theory. Like all the points passed on the slow train ride to their city, they saw quantum gravity as a sequence of milestones—from Rosenfeld, Bergmann, and Dirac's early work to the ADM method, the Wheeler-DeWitt equation, field theory in curved space-time, the spinor formulation of general relativity, and so on. The ultimate goal was to find some common ground in which general relativity and quantum theory could converse, but this could be a long journey indeed.

This meeting place need not be space-time itself, nor even a geometry where physical distances can be measured, as Bergmann emphasized. Bergmann liked to quote Einstein, who said that one should not "assume that distances are entities of a special kind, different from other physical quantities." In other words, one should not "reduce physics to geometry."[26]

When Bergmann stepped down, he took on a postretirement position at New York University, where his friend Engelbert Schucking kindly arranged office space for him. The Syracuse group continued to maintain Bergmann's philosophy of trying to find descriptions of general relativity that do not depend on a background metric (rule sheet indicating distances between space-time positions). Space-time distances, they felt, should be subservient to quantities more readily manipulated in quantum theory.

In 1986, physicists Amitaba Sen and Abhay Ashtekar developed a novel formulation of general relativity, based on its underlying connections, rather than on its space-time structure. In standard general relativity, connections indicate how space-time curvature affects the parallel transport of vectors (moving parallel to themselves along two different paths), as in the case we considered of Marius and Darius each walking to the North Pole (described in chapter 4). In the Sen-Ashtekar formalism, a special kind of connection becomes the fundamental physical variable, rendering Einstein's equations more amenable for quantization. Connections have properties similar to the $SU(2)$ non-Abelian gauge group, so this reformulation shapes general relativity into something like a Yang-Mills type of theory.

Seizing upon this new formalism, physicists Lee Smolin and Carlo Rovelli, then at Yale, set out to find solutions of the Wheeler-DeWitt equation of quantum gravity. Prior to their work, few solutions had been found to this equation, leading some researchers to conclude that its introduction had been a blind alley. Remarkably, Smolin and Rovelli found that they could solve the equation in terms of systems of loops, related in turn to the connections.

Smolin moved on to Syracuse, joining a new group of "loop theory" relativists headed by Ashtekar. Rovelli spent some time with the group as well, before becoming a professor at the University of Pittsburgh. The Syracuse collaboration expanded the loop formalism into a microscopic quantum description of gravity, showing how the universe could be fashioned from foamlike building blocks. Their model helped realize an earlier notion by Wheeler that on the quantum level geometry is as frothy as the sea on a stormy day. It also helped advance Bergmann's goal of separating physics from specific notions of space and time.

Drawing acclaim for their work, the loop group would branch out to other institutions. In 1994, Ashtekar became director of a new Center for Gravitational Physics and Geometry at Penn State University, bringing Smolin with him. Then in 2001, Smolin was appointed to a leadership role at the Perimeter Institute in Waterloo, Canada.

Bergmann followed these developments in quantum gravity with great interest, participating as much as he could in his later years. After Margot passed away, his health declined considerably. They had been best friends for more than sixty-five years, taking a life's journey together from the Black Forest in Germany to Riverside Park in New York.

In 2002, shortly before Bergmann died, the American Physical Society jointly awarded Wheeler and him its inaugural Einstein Prize for gravitational physics. Wheeler called Bergmann's son's house, where he was staying, and left a warm congratulatory message. Sadly, Bergmann passed away before learning about Wheeler's heartfelt gesture.

The Dynamic Dual

To an outsider during the final decade of the twentieth century, the two leading models of quantum gravity—string theory and loop theory—would sound strikingly similar. After all, both purport to be models of gravity involving minute, one-dimensional cords.

Yet on a fundamental level, the models have significant differences. String theory attempts to be a Theory of Everything, uniting gravity with its brethren forces. Loop theory makes no such claim. It is exclusively a way of breaking up general relativity into bite-sized pieces quantum theory can digest.

Moreover, string theory is a Kaluza-Klein theory, making sense only in higher-dimensional realms. It involves hypothetical particles and symmetries experimentalists have struggled in vain to detect. Loop theory, on the other hand, is a clever rewrite of Einstein's standard four-dimensional approach. For better or worse, it therefore does not stretch the imagination as much.

Finally, string theory relies on perturbative methods for calculating physical quantities. This involves adding up special Feynman diagrams. For these techniques to work, one must assume that interaction energies are relatively low. Loop theory, in contrast, is a nonperturbative model that applies particularly to high energies, such as conditions during the early universe. Its low energy limit is simply general relativity itself.

Throughout the history of theoretical physics various problems have called for perturbative versus nonperturbative methods. Physicists often use perturbative methods when exact solutions to equations are difficult or impossible to obtain. These involve building up an answer, step by step, by making a series of increasingly precise approximations.

For instance, suppose one wishes to estimate the volume of a tree. One might first measure the height and width of the trunk and estimate its bulk. This would provide a reasonable first guess. Then one might add to this figure the approximate volumes of the major branches, then the minor branches, and so on. Through this perturbative approach one would then home in on the right answer.

Nonperturbative techniques, on the other hand, involve using physical or mathematical principles to find exact solutions to problems. Often these make use of inherent symmetries or conservation laws that help simplify complex systems of equations. A nonperturbative way of determining the volume of a tree, for example, would be to uproot it, submerge it completely in water (by weighing it down), then measure the amount of water displaced. According to the laws of physics, this quantity must equal the tree's volume. If the tree is large, however, carrying out this method might be easier said than done.

Gradually, during the late 1980s and 1990s, a nonperturbative extension of string theory would emerge. This novel approach would encompass not just one-dimensional strings, but also two- and higher-dimensional objects, known as membranes, pulsing within an eleven-dimensional universe. These floppy entities have special relationships to one another—and to one-dimensional strings themselves—dictated by particular mathematical rules. Because these rules are exact, they lend a measure of unity and order to the superstring world.

The basic idea that particles might be membranes rather than points dates back to a proposal by Dirac in the 1960s. Until string theory became popular in the 1980s, few had even heard of Dirac's suggestion. Then, once one-dimensional objects came into vogue, the voices suggesting the incorporation of two- and higher-dimensional entities began to speak increasingly louder. (One must be careful to distinguish between the dimensionality of the objects and the dimensionality of the space itself; these are generally different.)

The brany breakthrough finally came in 1986 when University of Texas researchers James Hughes, Jun Liu, and Joseph Polchinski fashioned the first supersymmetric theory of extended objects. These supermembranes, as they were called, embraced vibrating, multidimensional shapes behaving like fermions as well as bosons. Then in 1987, physicists Eric Bergshoeff of the University of Groningen, Ergin Sezgin of the ICTP, and Paul Townsend of the University of Cambridge devised a model of two-branes (two-dimensional supermembranes) wiggling and jiggling in an eleven-dimensional universe with the features of supergravity. Townsend would designate p-dimensional branes, where "p" is any specific number of dimensions, by the appellation p-branes.

The same year, British physicists Michael J. Duff, Paul Howe, and Kellogg Stelle joined Japanese physicist Takeo Inami in proving that two-branes, wrapped tightly around a circle like paper towels around a tube, bear properties similar to superstrings. In fact, two-branes in eleven dimensions look just like strings in ten dimensions. All these results suggested close connections between membrane theory, string theory, and supergravity.

Also around the same time, Duff, Townsend, and others discovered remarkable dualities between various membrane and string models. A duality is a property of theoretical physics in which exchanging various parameters produces similar results. It can offer curious connections between models of vastly different properties.

For example, consider a five-year-old and a ninety-five-year-old walking up to the box office of a movie theater. A fifty-year-old, standing behind them, notices that they are both admitted for half price, but that he has to pay the full amount. Somehow, a duality between low age and high age places both of them into a discount category that middle age fails to provide.

Duff showed, for example, that a type of duality offered a linkage between one-dimensional strings and five-dimensional membranes, or five-branes. This duality involved exchanging the number of dimensions for the electric and magnetic parts of the field equations. Then in 1990, Strominger proved that five-branes could emerge from the heterotic string equations themselves as bulky companions to weakling strings (weak, that is, in terms of interactions with other strings). Somehow, on the beach of higher dimensions, every scrawny one-dimensional object is accompanied by a massive five-dimensional body—presumably to protect it from getting hyperspace sand kicked in its face. Moreover, in a form of divine justice offered by duality, the measlier the string, the beefier the five-brane. Furthermore, the converse is also true, enabling pumped-up strings to best their dimensionally gifted bodyguards. These relationships provided even stronger evidence that the fates of strings and membranes are tightly intertwined.

Yet, as Duff relates, most string theorists of the time wanted no part of membrane theory. "One string theorist I know would literally cover up his ears whenever the word 'membrane' was mentioned within his earshot!" says Duff. "Indeed I used to chide my more conservative string theory colleagues by accusing them of being unable to utter the M-word."[27]

Just as it took the blessings of such a respected figure as Witten for superstring theory to gain respect, it required the same physicist's santification for membrane theory to win acceptance. Certain duality relationships convinced him that all five superstring models could be united under a brane umbrella. He wasted no time in spreading the good word.

Mother of All Theories

In February 1995, at a University of Southern California superstring conference, Witten stepped up to the podium brimming with excitement. He heralded the dawning of a new unification approach, which he dubbed 'M-theory.' This has also become known as the second superstring revolution. Drawing on the work of Duff, Townsend, Strominger, and others, he had found a way of linking all the known types of strings in a single approach. No longer were there five competing models; now there was only one.

Witten forged his connections between the five categories of strings by use of two different types of duality, called S-duality and T-duality. S-duality, based on a conjecture by physicists Claus Montonen and David Olive, as further developed by Ashoke Sen and others, involves exchanging a weak coupling constant (indicator of interaction strength) for a strong one. T-duality, grounded in early superstring calculations, pertains to the replacement of a small compactification radius with an enormous one. That is, it trades Klein's tiny circle for one of astronomical proportions. Combined into a "duality of dualities," these transformations furnish the magic touch for one type of string to resemble another—even for a theory with only closed strings to become an open/closed mixture.

Witten was uncharacteristically elusive when explaining the term 'M-theory.' " 'M' stands for 'Magical,' 'Mystery,' or 'Membrane,' according to taste," he said. Others soon chimed in that it was the "Mother of All Theories." Duff was disappointed that Witten didn't directly call it "Membrane theory." He saw this ambiguity as a leftover from the days when string people disregarded him and his colleagues. "That the current theory ended up being called M-theory rather than Membrane theory was thus something of a Pyrrhic victory," suggested Duff.[28]

The elegance of M-theory impressed even many of those who had remained outside the fray when it came to strings. Its unity of vision and nonperturbative explanations allured those had been put off by the redundancies and approximations of the original superstring model. Many loop theorists admired M-theory's rigorous results, setting their sights on uniting it with loop quantum gravity into a single explanation for gravitation on all scales. The supergravity crowd was pleased because the low-energy limit of M-theory turned out to be eleven-dimensional supergravity. They felt that the new model vindicated their holding out for an eleven-dimensional approach. Even Glashow admitted that the string and membrane communities had made more progress than more traditional gauge theorists during the same period.

A 1996 paper by Witten and Princeton postdoctoral fellow Petr Horava splendidly mapped out the topography of M-theory, particularly with regard to the Heterotic-E type string. Like audiophiles eager to test new speakers, Horava and Witten cranked up the value of the string's coupling constant. As they dialed up the strength higher and higher, an amazing thing started to happen to the string. From a combination of dualities, it grew thicker and thicker, not along any direction in its own space, but rather into an extra perpendicular dimension. It had evolved into a two-brane. The result was just like A Square from Flatland rising out of his plane and discovering a whole new higher-dimensional world!

Horava and Witten's novel topography soon acquired the catchy designation 'brane world.' The name derived from the special layout of their model. The Princeton researchers found that their cosmic blueprint divided neatly into three distinct spatial sectors. First, there was ordinary three-dimensional space, which they rechristened the 'three-brane.' Second, there was the extra dimension that arose from duality principles. Bounded by the three-brane, this spanned a four-dimensional domain called the 'bulk.' Finally, there was the six-dimensional compactified region, a miniscule Calabi-Yau shape twisted up beyond all possible detection. This Calabi-Yau sector housed the symmetries of the standard particle model. Thus there were ten spatial dimensions in total: four large and six compact. Add time to the mix, and Horava and Witten offered a road map for an eleven-dimensional "brane world" universe.

Curiously, closed and open strings have different levels of access to this cosmos. While closed strings can travel anywhere they want, including the bulk, open strings are confined such that their ends must always remain in the same three-brane. Thus open strings have one less degree of freedom than closed strings, significantly changing their dynamics.

This restriction was discovered by Polchinski, who used the power of T-duality to prove that open strings and so-called Dirichlet branes (D-branes, for short) have a symbiotic relationship. Like koalas and eucalyptus trees, one cannot find the former without the latter. The three-brane in the Horava-Witten model thus represents an example of a three-dimensional D-brane. Open strings stick to it, while closed strings are free to wander off.

Gravitons, the carriers of gravitation, are modeled by closed strings, while W-bosons, Z-bosons, photons, and gluons, the carriers of the other natural forces, are represented by open strings. This suggested that the difference in behavior between closed and open strings could well explain a long-standing riddle: why is gravity so much weaker than the other forces?

Soon the race was on to develop credible brane worlds, ones that would offer physically realistic descriptions of nature's varied range

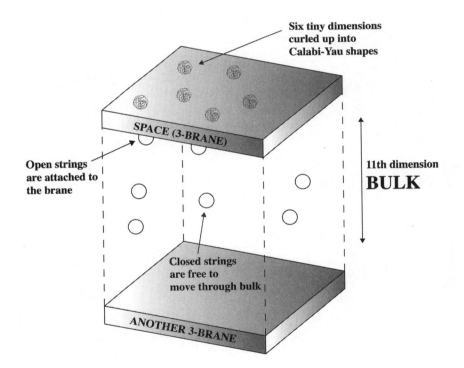

Depicted here is an eleven-dimensional brane-world scenario. One of the branes in this "sandwich" encapsulates the three ordinary spatial dimensions. It is a Dirichlet brane, or D-brane, meaning open strings attached to it cannot escape. Six of the extra spatial dimensions are curled up into minute Calabi-Yau shapes. In addition to time, that leaves one extra dimension, potentially large enough to detect. Only gravitons, being closed strings, are free to travel along this extra dimension, passing through a nether region known as the bulk.

of interactions. Because these worlds would have an extra noncompact dimension, researchers realized that this added feature might even be testable. While experimentalists savored the notion of examining the possibility of an additional dimension, theorists clamored to make sure their models would meet reasonable specifications. Tinkering with various physical and mathematical parameters, they set out to design a brane world suitable to represent our own world.

CHAPTER 12

Brane Worlds and Parallel Universes

One theory to rule them all,
One theory to find them,
One theory to quantize them all,
And in the darkness bind them,
In the Land of Brane where the shadows lie.

—With apologies to J. R. R. Tolkien

Brane Bank

In a rusty brick building just off the center of Harvard's leafy campus, young physicists are plotting out extraordinary new ways to describe nature. Jefferson Physical Laboratory, with its functional, late nineteenth-century architecture, presents itself as if it should be a traditional scientific enclave, full of glass cases, oscilloscopes, bunsen burners, and the like. It dates back to the era when the theories composed in one part of the building—typically the highest floors—could possibly be tested in another—typically the basement. Indeed, the west wing houses a special tower designed to measure how the rotation of the Earth causes dropping objects to deflect eastward. In the 1960s, the physicist Robert Pound used the same vertical shaft to measure the gravitational redshift of light and compare its value with Einstein's predictions.

Such experimental verification on site is not very likely these days, given the extraordinary energies required to test modern field theories. Nevertheless, the sacred scheme has remained, and the offices of the high-energy theory group are on the top floor of the building, closest to the heavens and farthest from the Massachusetts soil.

In a cozy arrangement Veblen would have favored, the theory group's quarters is a circle of offices surrounding a fishbowl seminar room. An extravagance of blackboards reaches from the carpet to the high ceilings, taking up all the available wall space. Rubbed clean every morning, by midday they are filled with a plethora of squiggles, loops, and equations. If these become too crowded, little nooks and crannies, each with their own blackboards and chairs, provide additional places for students to impress their advisers.

Glashow once vowed to keep string theory out of Harvard. Now to reach his office one must pass through a portal bordered by two of the best-known string and brane theorists: Nima Arkani-Hamed and Lisa Randall. Situated on either side of the entrance to the center, they represent independent, radically innovative ideas for modeling gravitation in the universe. The brane world notions of these scientific superstars have generated more buzz than virtually any other theoretical papers in recent years.

Just a Millimeter Away

With his long brown hair parted in the middle, youthful good looks, and charismatic style, Arkani-Hamed appears every bit the rock star. He would seem more at home with the casual crowds enjoying concerts on Harvard Square than with the starched shirts attending official university functions. Wielding chalk instead of a pick, he does draw quite a following in his own way. Rather than soaking in the strains of a guitar, the graduate students that gather around him tune into his riffs on microscopic vibrations floating through unseen worlds.

In his thirty-something years, Arkani-Hamed has passed through quite a few worlds himself. He was born in Houston to two Iranian physicists on leave from their university positions. Shortly thereafter, his family returned to Iran. Then came the revolution, which caused them grave political difficulties because of their high scientific status and Western connections. Consequently, they were forced to flee the country on horseback.[1] Eventually, they ended up in Toronto, where his father received a new appointment.

Safe in a Canadian high school, Nima became fascinated by the family profession. Enrolling at the University of Toronto, he majored in both math and physics. Then, heading off to Berkeley, he obtained a PhD in theoretical physics. This led him to a postdoctoral fellowship at the Stanford Linear Accelerator Center (SLAC) theoretical group, beginning in 1997. As someone who revels in scientific discussion, he started to look around for suitable collaborative projects.

As luck would have it, Savas Dimopoulos, a Stanford physicist esteemed for his pivotal work in supersymmetry, was also seeking new research opportunities. Dimopoulos grew up in Athens, Greece, then completed his Ph.D. at the University of Chicago. In the 1980s, along with Howard Georgi, he developed supersymmetric versions of the standard model and Grand Unified Theories. Although their theories made testable predictions, current accelerators and detectors seemed inadequate to rule them in or out. Therefore, while waiting for the next generation of accelerators, Dimopoulos was open to consider alternative hypotheses, especially those that experimentalists could test more easily. "I think it is our job as physicists to make predictions that are falsifiable by experiment," Dimopoulos has said.[2]

His interest in testable conjectures led him to develop, along with Arkani-Hamed, as well as physicists Georgi (Gia) Dvali and Ignatius Antoniadas, a bold answer to a long-standing physical problem. One of the greatest mysteries about the natural forces is why gravity is so much weaker than the others. For example, a pocket comb rubbed with a piece of cloth can sustain enough static electrical force to outmatch the whole earth's gravitational attraction and pick up bits of paper from a table. Dirac and Jordan each tried to address this issue—known as the "hierarchy problem"—in their theories of a varying gravitational constant. The Stanford collaborators took another tact, proposing that the difference stems from gravitational leakage into a large extra dimension.

In several influential papers, one with Antoniadas and the others without, the researchers described the architecture of a brane world that would explain gravity's relative weakness. The first component of their model is our own familiar space. As in the Horava-Witten model, they identified this as a three-dimensional D-brane, otherwise known as a three-brane. Like selective flypaper, this three-brane is sticky for open strings, while letting closed strings get away. Almost every particle with which we are familiar is trapped forever on our brane. The principal exceptions are the gravitons. As closed strings, they are free to venture off our brane into another dimension.

If this were the complete picture, then gravity would be *infinitely* weaker than the other interactions. Like untrained pets, all the gravitons would simply wander off forever. There would be no chance whatsoever for two massive objects in our own space to experience a mutual gravitational attraction. What, then, keeps the gravitons fenced in enough for interactions to happen at all?

The answer, as the Stanford team proposed, is to picture a second three-brane less than one millimeter (⅒₅ of an inch) away from our own, along an extra dimension. This fences the gravitons into a finite region—a bulk between the two branes—localizing gravity, but allowing it just enough wandering room to account for its relative weakness.

The distinction between gravitons trapped on our three-brane, gravitons confined to a small but finite bulk, and gravitons completely free to wander off is akin to the difference between having a wolf in one's home, in one's backyard, or totally unfettered. Stuck inside a house, a wolf would have the most impact, destroying furniture, terrorizing guests, and demolishing one's food. Fully liberated, the beast could roam the open wilderness, possibly having little to no contact with people. The middling scenario would be a wolf stalking through a small fenced-in garden. It would create more mayhem than if it were free, but certainly far less than if it were inside. Similarly, gravitation limited to a one-millimeter-wide bulk has some strength, but far less than the brane-confined forces.

The scheme crafted by Arkani-Hamed, Dimopoulos, and Dvali (with suggestions from Antoniadas) became known as the "large extra dimension" scenario, or sometimes just ADD for their initials. Normally we don't think of a millimeter as large, but it certainly is compared to the minute scales proposed for previous compactification schemes.

One would think that the addition of a large extra dimension would radically change the laws of physics. For example, Ehrenfest predicted that more than three spatial dimensions would alter the familiar inverse-squared form of gravity. On astronomical scales, this would fail to produce stable planetary orbits.

However, at the point the ADD model was introduced (the late 1990s), gravitation had not been tested down to the millimeter level. No one knew if the law of gravity had exactly the same form on the tiniest scales. This left the door open for experimental verification of the theory. For the first time in their long history, extra-dimensional models could be placed on the examining table. Rejoicing at their newfound opportunities, experimentalists immediately set out to work, designing tests to explore the submillimeter structure of gravity.

The Neighboring Darkness

Walking down a lonely road at night, one is startled to come across an unexpected nearby presence. An accidental close encounter with a stranger can lead to shivers down one's spine, even if the passerby is simply an innocent pedestrian walking his dog. Now imagine feeling the hot breath of an alien entity less than one millimeter away. Such is the proximity of the parallel universe envisioned by Arkani-Hamed and his collaborators. Creepily close, it would be separated from us by only the thinnest curtain. Yet, for all intents and purposes, that barrier would be impenetrable, except for the force of gravity. Like the Berlin Wall during the Cold War years, it would keep us from contacting what lies on the other side.

Strangely, astronomers come across unexplained ghostly forms in their ordinary research. For years, they have realized that the overwhelming majority of the universe is filled with various types of invisible matter. This material makes its presence known only through its gravitational influence.

For example, the peripheral stars in galaxies swirl around much faster than they ought to, if all the matter in galaxies could be mapped out. Some sizable chunk is missing, making its influence known only through its gravitational pull. Astronomers believe that a small portion of this unseen substance lies in very faint stellar bodies, known by the acronym MACHOs (Massive Compact Halo Objects). Another segment pertains to the combined mass of neutrinos, particles notoriously hard to detect. The vast remainder of the material is of unknown origin.

In 2003, the Wilkinson Microwave Anisotropy Probe (WMAP) produced an extremely detailed map of the sky's microwave radiation. NASA designed this special satellite and sent it into orbit to record minute temperature differences in the cosmic background radiation left over from the Big Bang. In tandem with other measuring instruments, this probe has offered a glimpse of the content and distribution of material in the universe.

The results of the new cosmological measurements have confirmed the puzzling composition of the universe. WMAP indicates that only 4 percent of the cosmos consists of conventional baryonic matter, such as protons and neutrons. About 23 percent lies in the form of so-called cold dark matter. This is low-temperature, nonbaryonic material that interacts by means of gravity, not light (or at least not very easily with light). The remaining 73 percent is something called dark energy—a kind of antigravity that causes the universe to accelerate outward. One of the greatest mysteries of modern cosmology is the nature of these hidden substances.

Arkani-Hamed, Dimopoulos, Dvali, and Stanford researcher Nemanja Kaloper have postulated a particular brane world scenario that could possibly resolve the dark matter dilemma. Their proposal, called the Manyfold Universe envisions a single brane folded up along a higher dimension into something like a fan or an accordion. One of the folds corresponds to our own region of space, the others to regions extraordinarily far away. Filling the creases is the bulk, the razor-thin realm forbidden to all but gravitons.

Here is how the Manyfold Universe model could possibly explain dark matter. Suppose astronomers point their telescopes at a particular galaxy. Though the light from that region of space stems from stars and gases on our own fold, perhaps they would indirectly observe the gravitational effects of materials on other folds. As in the Indonesian puppet shows described in chapter 1, maybe they would be detecting the gravitational shadows of parallel worlds operating behind the scenes of visible reality. The astronomers might then

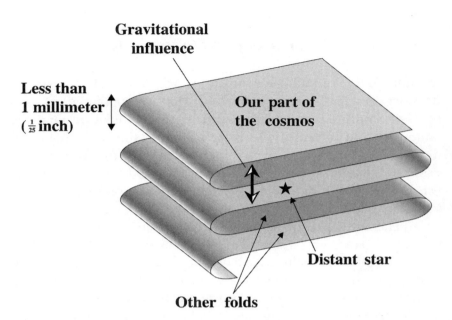

Shown is the topography of the Manyfold Universe. In this model, our brane is folded like a blanket. Only gravity can escape from the folds and pass freely through the bulk. Conceivably the gravitational attraction of a distant star, several folds away, could make itself known by its influence on objects in our fold. The light from the star, however, would be confined to the brane and would not exude through at the same location. Therefore we'd perceive the star's unseen influence as dark matter.

credit these unseen influences to MACHOs or other types of murky substances, but in reality they'd be due to extra dimensions.

For example, consider a star shining billions of light years (that is, many billions of trillions of miles) away. Its light would take billions of years to reach us. According to standard physics, its gravitational influence would take the same amount of time to arrive. However, in the Manyfold Universe theory, if it happens to lie on a separate fold from ours, its gravitons could arrive via an interbrane short cut. They could interact with neighboring material, distorting its trajectory through space. They could even reshape the paths of incoming radiation. In short, the distant star could exert a ghostly effect on our own region, manifesting itself as dark matter.

Deconstructing Dimensions

After his postdoctoral fellowship at SLAC, Arkani-Hamed was at Berkeley for a few years. Then in 2001, Harvard appointed him to a professorship. Along with Lisa Randall's appointment around the same time and Andrew Strominger's in 1998, this was a bold new direction for Harvard's physics department. Harvard was vying with Princeton, Stanford, Caltech, Santa Barbara, and other more established string and M-theory centers to become one of America's leading brane banks.

At Harvard, Arkani-Hamed has decided to take a different direction in his research. Following the maxim that a collaborator of one's collaborator would make a good collaborator, he and Howard Georgi have been diligently working together. Besides, they are just down the hall from each other, saving greatly on travel costs.

Georgi is extremely open to collaborative discussions, as indicated by a prominent sign in front of his office that reads, "If the door is open, please walk in and have a seat." Under the message is another that says, "P.S. That's right, don't knock. Just walk right in." Below that is "P.P.S. I really mean it." A nearby bust of Einstein gazes intensely upon all who approach, making sure they follow these directions.

The new joint project, along with Boston University theorist Andrew Cohen, is called Deconstructing Dimensions. In literary theory, deconstruction is a movement started in the 1960s by French philosopher Jacques Derrida that involves a special method for analyzing texts. The term has since passed into popular usage to designate ways of breaking down bodies of thought into their fundamental aspects. In this spirit, Arkani-Hamed, Cohen, and Georgi hope to distill the very basis of physical dimensions from completely dimensionless constructs. (A second group from Fermilab has published a competing proposal along similar lines.)

Ironically, their approach brings M-theory more in line with the Syracuse program—as expressed by Bergmann and others—of transcending the framework of space-time geometry. In performing this feat, it would make the idea of a universal beginning even more mysterious. In the most radical version of this hypothesis, not only would the universe start off as tremendously compact object, it would begin as a literal point, with all of its dimensions produced dynamically.

It's hard enough to picture extra dimensions, let alone a world with *no* dimensions. In Abbott's *Flatland,* the protagonist dreams about Lineland, a realm with only a single dimension. He meets the king, whose movements are restricted to only left or right, and whose vision is confined to single points on either side. The king's perception of the world is thereby severely limited. Imagine the solipsistic existence of being in an entire cosmos with no spatial motions whatsoever. Our lives depend so much on movement and change that such a dimensionless world is virtually inconceivable.

The dimension-creating process proposed by the Harvard and Fermilab teams somewhat resembles the evaporation of a block of dry ice, only with temperatures falling instead of rising. In dry ice (or any solid for that matter), the molecules have relatively little freedom to move. Assembled in a lattice, they can exhibit only slight motions around their average positions. Now imagine if the molecules could not move at all in any direction (in conventional physics such a state, called absolute zero, is impossible). Effectively, they would be trapped in their own points, and would never know that dimensions exist. Under room-temperature conditions, however, the dry ice would soon become gaseous carbon dioxide. The molecules would then be able to travel in any direction, giving them full awareness of the three-dimensional world. Thus in some sense they would be gaining dimensions.

In the Deconstructing Dimensions model, lowering the universal temperature rather than raising it would precipitate such a phase transition. In the beginning of time, no forces would exist and hence no movements would be possible. As the cosmos cooled down it would transform from its initial state with no dimensions into a lower-energy state that has dimensions. The dimensions would appear with the creation of forces that allow for motion between separate points. Hence space would be liberated from its rigid "block of ice."

A more conservative version of the model envisions the universe starting out as ordinary four-dimensional space-time. In that case, only the extra dimensions would be produced dynamically. As temperatures lowered, new interactions would emerge that would allow for motion in higher dimensions. In this manner, conventional gauge theory of the sort pioneered by Georgi and his generation could magically transform itself into a large extra-dimensional approach of the type advocated by Arkani-Hamed and his colleagues.

Hey, Maldacena!

One of the emerging themes of early twenty-first-century physics has been the establishment of closer connections between traditional high-energy methods and more radical Kaluza-Klein-type approaches, including brane and string theories. The collaborative efforts of Arkani-Hamed, Cohen, and Georgi provide only one example of the growing set of strands knitting together the two communities.

Greatly stimulating this dialogue is a rather powerful conjecture put forth by the young Argentine physicist Juan Maldacena. Since its introduction in the late 1990s, Maldacena's proposal has served as a kind of Rosetta Stone, allowing valuable translation from the language of multidimensional M-theory to that of four-dimensional gauge theory, and vice versa.

When physicists at a Santa Barbara string conference first learned about Maldacena's magnificent bridging of the two systems of thought, they literally danced in the aisles. In the manner of a hip-hop DJ, string master Jeffrey Harvey (of Princeton quartet fame) began to rap out satirical lines, set to the tune of the popular number "The Macarena." As Maxwell did in his scientific poetry, Harvey cleverly versified the new discovery. Meanwhile, hundreds of string theorists lined up and performed the elaborate gestures associated with the dance. The lyrics began:

> You start with the brane and the brane is B.P.S.
> Then you go near the brane and the space is A.D.S.
> Who knows what it means—I don't, I confess
> Ehhhh! Maldacena![3]

In Harvey's parody, the term B.P.S. (Bogomol'nyi-Prasad-Sommerfield) refers to a special supersymmetric condition for branes. A.D.S. stands for anti-deSitter space, a particular variation of

Einsteinian general relativity that includes a negative cosmological constant (antigravity term). Anti-deSitter cosmologies tend to slow down over time, as opposed to De Sitter universes that tend to speed up. Harvey's mention of a B.P.S. brane and A.D.S. space present two of the requirements for Maldacena's theory. The parody goes on to describe other aspects of the approach.

Expressed in scientific prose rather than danceable verse, Maldacena's conjecture involves a mathematical connection between the properties of strings in a saddle-shaped five-dimensional bulk and the structure of standard quantum field theory in the four-dimensional region bounding that space. Etched on the exterior of the M-theory universe is the gauge theory for a supersymmetric Yang-Mills model. This interests high-energy physicists in general, not just string theorists. It is as if participants at a string conference tapped out messages on the windows for their more traditional colleagues standing outside. "We're doing something important inside, so here's an exterior glance to give you a sense of the applications," the messages would seem to say.

Maldacena's proposal is closely related to the *holographic principle*, a hypothesis first advanced by Geraldus 't Hooft and Leonard Susskind. In this context, holography involves a relationship between the information content on the boundary and interior of a spatial region. It is an extrapolation of what is known about black holes, the highly compact stellar remnants christened by John Wheeler. Scientists characterize black holes not by what lies within, but rather by their exterior properties. For example, when black holes gobble up material, their bounding surfaces grow to encompass greater area. This area is proportional to the entropy, or amount of disorder, within the black hole. The holographic principle suggests that this is true for the universe in general. That is, to know the cosmos, one needs only to understand its outer limits.

Describing a closed volume by means of its exterior alone constitutes dialing the number of dimensions down one notch. Therefore, to know a three-dimensional bubble one need only examine its thin surface, and to understand a five-dimensional bulk, its four-dimensional brane wall (a three-brane plus time) might nicely suffice. Hence, Maldacena's magic feat. Or as Harvey puts this:

Who says they're the same—holographic he contends?
Ehhhh! Maldacena![4]

The holographic principle and Maldacena's conjecture have provided ample discussion at string theory conferences, whether in mathematical or musical form. But these conferences aren't all song and dance. In between blocks of sessions, it's time for communing with nature and asking her privately if any of these hypotheses are correct. This dialogue with the great beyond usually involves some kind of outdoor sport.

Warped Terrain

When Bohr needed some peace of mind to think about the fundamental contradictions of quantum physics, he used to go hiking or skiing. Often Klein or his other assistants would join him. Today one of the most popular sports among theorists involves ascending much steeper and more rugged terrain: rock climbing, that is.

It takes a special nerve to hoist oneself up a sheer cliff, handhold by handhold and foothold by foothold, while praying for the solidity of the precipice to which one's limbs and fate are attached. After battling wind, sun, and vertigo, vying with even the most intractable set of equations seems like child's play by comparison. What better antidote to the fear of presenting papers in front of critical audiences than facing the ultimate terror that one's next step could be one's last.

Clinging to a rock face, one might sympathize with the plight of open strings. Their ends are forever attached to D-branes, while their closed-string neighbors are free to flitter from wall to wall through the grander bulk. No wonder rock climbing and mountaineering are such popular activities for M-theorists. Aficionados include Steven Giddings of Santa Barbara, who habitually conquers peaks in the Rockies and Sierras; Steven Gubser of Caltech, who has scaled craggy formations in the Grand Tetons; Georgi Dvali, now at New York University, who has trekked through the Caucasus Mountains; and Burt Ovrut of the University of Pennsylvania, who has climbed the monumental face of El Capitan in Yosemite.

Ovrut finds close connections between the skills needed for theoretical physics and technical climbing. "In one you're learning something abstract about nature and how it works," he says. "In the other you're following these beautiful cracks of rock. Both are elegant things to be understood and worked out."[5]

For Lisa Randall, another keen climber, her mountaineering skills are matched only by her adeptness at conquering other kinds of barriers. In an era where it is still unusual for women to hold professorships in theoretical physics, she has served as an outstanding role model for a new generation.

Born in Queens, New York, in 1962, Randall grew up in a family in which mathematical talent was the rule rather than the exception. By a young age, she had established her extraordinary gift for problem solving. Attending Stuyvesant, the premier academic high school in Lower Manhattan, she joined its award-winning math team, where she rose to the rank of co-captain. Known as a springboard for successful careers, the group had a regimen that included working out mind-bending problems virtually every day. Brian Greene, who ended up a well-known string theorist and author, was on the same team until he became too busy with other pursuits. Lisa's younger sister Dana was on the math team as well and ultimately became a professor at Georgia Tech.

In 1980, the year of her graduation from Stuyvesant, Randall was a winner of the prestigious Westinghouse Science Talent Search, a national contest for identifying gifted young scientists. She went on to Harvard, where she received both her B.A. and Ph.D. Her thesis supervisor was none other than Howard Georgi, who mentored her in the nuances of modern supersymmetric field theory. After appointments at Princeton and MIT and awards including a Sloan Fellowship, she returned to her alma mater in 2001 as professor of physics.

By then, Randall had become well known for an innovative approach to higher dimensions developed jointly with Raman Sundrum. With its clever notion of using the warping of the bulk to localize gravity near the brane, the Randall-Sundrum model offered an intriguing alternative to the ADD model. Randall and Sundrum's papers have stimulated intense dialogue between string theorists and cosmologists exploring the relationship between their two disciplines.

Sundrum was born in Madras, India, in 1964. Unlike his theory, he has hardly led a localized existence. In 1965, his family moved to the United States, then returned to India in 1970. They remained there for three years before they were off to Canberra, Australia. Sundrum completed his undergraduate degree at the University of Sydney before flying over to Yale and carrying out his graduate studies. His Ph.D. adviser was cosmologist Lawrence Krauss, who has subsequently become a leading popularizer of science. Currently Sundrum is a professor at Johns Hopkins.

Unlike some of his peers, Sundrum has no interest in rock climbing or mountaineering. Preferring quieter pursuits such as reading, he has long been fascinated by philosophical questions in science. Probing the nature of artificial intelligence and examining the mystery of evolution have been some of his pet topics, when he is not exploring higher dimensions.

One of the baffling questions that drew Sundrum into the world of M-theory concerns the size of the cosmological constant. The Casimir effect predicts that even empty space must contain a certain amount of energy. This vacuum energy manifests itself in Einstein's equations as an extra antigravity term. The presence of a significant cosmological constant leads, in turn, to a prediction that the universe's expansion is highly accelerated. In other words, theoretical physics ordains that the cosmos should be blowing itself apart at an ever-hastening rate. As recent measurements have shown, the universe is indeed accelerating, but at a far slower pace than theory suggests.

In an effort to explain the smallness of the cosmological constant, Sundrum teamed up with Randall and developed a brane world with substantially different features from the ADD approach. Instead of a flat, millimeter-sized extra dimension filling the gap between two branes, they took the Maldacena conjecture to heart and developed a saddle-shaped, anti-DeSitter universe surrounding a single brane. (Another paper they wrote together has two branes, making the designation of Randall-Sundrum model somewhat ambiguous; we focus here on their single-brane theory.) Borrowing a term from the *Star Trek* television series, a special tuning parameter called the "warp factor" governs the shape of the extra dimension.

The beauty of an anti-deSitter space is that it has a negative cosmological constant of its own. With luck, this precisely balances out the vacuum energy on the brane, leaving only a small residual effect. Depending on how fast astronomers observe the universe to be accelerating, one can adjust this balance accordingly.

Moreover, the negative curvature has a concentrating effect upon gravity. For all intents and purposes, it confines the gravitons to a basin very close to the brane. Like the waters of the Colorado River flowing through rocky gorges, they are trapped by the topography. Only at extremely high energies, beyond current detection, might they escape from such a "Grand Canyon." Hence, although gravity can leave the brane, making it weaker than the other forces, it can't spread out far enough to render itself insignificant.

Much to Randall and Sundrum's amazement, this focusing effect occurs for extra dimensions of any size at all. Even if the higher dimension is infinite, gravity can still take the form that it does in the real world. Though it is as vast as space itself, the extra direction need not reveal itself. Consequently, the researchers realized that they had found a viable alternative to compactification. Extra dimensions need not curl up; they can stretch out forever but still localize gravity.

Randall and Sundrum's proposal rocked the physics community like a new Beatles record. Overnight, scores of theorists began to dance to the new beat (not literally, though, as in Maldacena's case). As Sundrum recalls, "The alternative to compactification paper was just a real surprise, because there really was apparently no way around having to compactify up to that point in time. It was a big magic trick that was quite astonishing for most people including its authors. Also, it was very mysterious to have a secret, curved spacetime."[6]

The Infinite Reaches

The Randall-Sundrum model was not the first to envision infinite extra dimensions. As early as 1962, Purdue University researcher David W. Joseph suggested that familiar space-time lies in a "potential trough" (a dip in potential energy) surrounded by vast reaches of multidimensional space. Curiously, his relatively obscure writings on the subject made no reference to Kaluza-Klein theory. Another little-known paper, by Russian theorists Valery Rubakov and Misha Sha-poshnikov, was the first to propose that extra dimensions could dilute the power of gravity.

Then, in 1986, Cambridge physicists Gary Gibbons and David Wiltshire revisited the topic. They envisioned four-dimensional mem-branes (before they were nicknamed "branes") embedded within a higher-dimensional universe. The higher-dimensional regions would be quite inhospitable, consisting largely of black holes, leaving the four-dimensional substructures to be the tamer abodes of living beings. Somehow the laws of physics would guarantee that everything we observe would be confined to the latter rather than spread out through the former. Much to their disappointment, however, the Cambridge theorists could not find a way of localizing gravity.

Several years later, astrophysicist Paul Wesson of the University of Waterloo in Canada developed a classical (nonquantum) five-dimensional theory, equally vast in all directions. Resurrecting Ein-stein's notion that a unified theory ought to explain matter through geometry, Wesson modified the Kaluza-Klein hypothesis by remov-ing the requirement that the fifth dimension be compact. This cre-ated extra terms in Einstein's equations that he identified with the matter and energy content of the universe. Wesson has since estab-lished an international "5D Space-Time-Matter Consortium" to pro-mote discussion of Kaluza-Klein ideas, particularly versions modified to include an infinite extra dimension.

What these approaches have in common is a feeling that com-pactification of some dimensions but not others destroys the equity of nature. Why should some of reality's highways extend forever, while others are curled up into tight roundabouts? Which engineer mandated that length, width, and height should be straight and infi-nite, while other dimensions perform tiny loop-the-loops?

On the other hand, as Gibbons points out, the standard Kaluza-Klein compactified approach has certain advantages over infinite extra dimensions. "The original model is rather complete," says Gibbons. "It limits information. A single brane with an anti-deSitter background has potential difficulties with singularities. Information can come from outside the universe."[7]

Gibbons is concerned that a Randall-Sundrum universe could subvert the law of cause and effect. Events could occur at any time with origins beyond the space we see. Virtually anything could pop out. A shark could suddenly materialize in one's swimming pool because of some strange interplay between the bulk and brane. Our sense that we might someday understand the world in its entirety would become increasingly precarious. "I like to think of it as a biological picture," Gibbons says. "Like algae on a pond, someone can come in and destroy you at any moment."[8]

As for Randall and Sundrum, they have each moved on to other kinds of models. Maintaining an open mind, each has investigated a variety of approaches, ranging from more traditional compactified theories to those with different arrangements of branes. For instance, Randall is not sure if a one-brane or two-brane configuration would be ideal for describing the universe.

"They have different applications," she points out. "I have one student working on a two brane anti-deSitter space that reproduces four dimensions. We are still trying to see which theory is best." At this stage of the game, she says, "we just want to know what the possibilities are.[9]

When Universes Collide

Paul Steinhardt has been a longtime leader in the field of cosmology. He is best known as one of the codevelopers of the inflationary universe model. Inflation supplements the primordial Big Bang with a brief but explosive interval of rapid expansion. In doing so, it helps resolve certain long-standing dilemmas, such as why the universe appears relatively uniform in all directions. When space stretches out so quickly, all its blemishes get smoothed out, leaving it as flat as a pancake.

Recent astronomical findings have boded well for inflation. Satellite images of the cosmic microwave background radiation, such as those taken by WMAP, have supported its general story of how events panned out (albeit not its simplest version). This has given Alan Guth, Andrei Linde, and Andreas Albrecht—the other coinventors of the approach—much cause for celebration.

Though Steinhardt has similarly been pleased, he sees considerable danger in relying on one particular model. "When you get down only to a single competitor it's not always a healthy situation," he advises. "It's much better to have two or more competing models, forcing you to think more carefully about your theories, your predictions and the observations."[10]

For this reason, he set out with string theorists Bert Ovrut of Penn, Neil Turok of Cambridge, and Justin Khoury of Princeton to fashion an alternate explanation for cosmic uniformity based upon a brane world scenario. Putting their branes together, they developed a collision scenario that would mimic the energy spectrum produced by inflation. Two branes—one of them representing our own space and the other a rogue three-brane that happened to head in our direction—would smash into each other along an extra dimension. During this cosmic catastrophe, our brane would heat up like a wrecked vehicle. All this surplus energy would convert into hot matter and radiation, essentially producing the Big Bang conditions of the early universe without the initial singularity (moment of infinite density). Because the two colliding branes would be flat, the byproduct would essentially be uniform, mimicking the effect of inflation. Remarkably, the cosmic background radiation spectrum produced would also be similar to that of inflation. The researchers dubbed this model the Ekpyrotic Universe, named after the Greek Stoic belief in cosmic renewal through fire.

In a related proposal, called the Cyclic Universe, Steinhardt and Turok envision how the collision of two branes could create a cosmos that has no true beginning or end. Resurrecting ancient notions that everything experiences endless periods of creation and destruction, they show how a special brane world scenario could explain current astrophysical data without reference to a first or last moment of time. Periodically, the interaction of our brane and another would flood the universe with energy, wiping out all existing things (except perhaps for black holes) and planting the seeds for new galaxies. These galaxies would develop for billions of years, harboring various forms of life, until they are destroyed in a catastrophic "Big Crunch." The process would happen again and again for eternity. The authors feel that their new approach could reduce philosophical issues concerning the origin and fate of the cosmos, while echoing the predictive strengths of the Big Bang/inflationary epic. In addition, as they have devised it, their model includes an explanation of the recently discovered acceleration of the universe.

Though the Ekpyrotic and Cyclic proposals provide intriguing, dramatic alternatives to more conventional models, they have yet to win over many supporters. Steinhardt and his collaborators have found skepticism in both the cosmology and superstring communities. He is disappointed that the other codevelopers of inflation refuse to consider brane world approaches as a serious alternative.

Linde, for example, has been extremely dubious. In a talk in honor of renowned Cambridge theorist Stephen Hawking's sixtieth birthday, he delivered a point-by-point critique of both proposals. He pointed out a number of areas where, in his opinion, inflation made successful predictions that the brane world models couldn't match. Nevertheless, because it was a birthday celebration, Linde ended the lecture on a positive note. He congratulated the authors of the Ekpyrotic/Cyclic models for developing the best alternative to inflation he is aware of, proving that the inflationary model can beat even the boldest challengers.

In contrast to his expectations for his fellow cosmologists, Steinhardt has been less flummoxed by the reaction of bread-and-butter string theorists. He understands their belief that the time is not yet right for a cosmology based on branes. As he explains:

> There's a problem with string theory and cosmology in that string theory cannot deal with time dependence. They haven't gotten to that point yet. They are two fields with a very different character: cosmology, which is putting together a story, and string theory, which is a more formal mathematical subject. It's a cultural mismatch. So there's a skepticism on their part because they cannot say, "yes this is consistent with string theory." There's a certain level of discomfort. A lot of interest, but also a certain level of discomfort."

Randall agrees that brane cosmologies have a long way to go before they are endorsed by the string community. "They haven't been fully developed yet," she says.[11] Sundrum seconds this view, remarking, "There are just too many bells and whistles, twists and turns. You can add and subtract and change the results you get."[12] Nevertheless, each supports a continued dialogue between string theorists and cosmologists, hoping that someday a complete model of the universe will unite the very large with the very small.

Putting It to the Test

Cosmology would be only one way of testing the higher-dimensional hypothesis. A more traditional means of ascertaining whether a field theory is correct involves the use of particle accelerators. Researchers used such "atom smashers" to confirm many twentieth-century theories, including the notion that quarks and leptons come in families as well as the standard electroweak model.

If supersymmetric companion particles were found, that would offer a tremendous boost to superstring, supergravity, and M-theory approaches. Theorists hope that an enormous new accelerator at CERN, called the Large Hadron Collider (LHC), will deliver the energies required to produce hypothesized objects such as squarks, selectrons, and neutralinos. Scheduled to open in 2007, it will smash pairs of protons together at energies of 14 TeV—the equivalent of billions of batteries concentrated within a point trillions of times smaller than a flea. Racing around a 16-mile ring in opposite directions, each pair will be traveling extraordinarily close to the speed of light before crashing into each other to produce a blast of particles. The LHC will be considerably more powerful than even the most energetic machine operating today, the Tevatron at Fermilab.

In and of itself, however, the experimental confirmation of supersymmetry would not be enough to prove the existence of extra dimensions. Because there are four-dimensional as well as ten- or eleven-dimensional supersymmetric models, the discovery of supersymmetric particles might not provide enough data to distinguish the two. Unequivocal proof of dimensions beyond space and time would need to come through independent means.

The prospects of discovering extra dimensions would greatly depend on their size. Tightly curled-up tubes, as miniscule as the Planck scale, could be probed only indirectly. In contrast, large extra dimensions of the type proposed by Arkani-Hamed, Dimopoulos, and Dvali would be far easier to examine. With their submillimeter-sized girth (less than $\frac{1}{25}$ inch), they would produce unmistakable discrepancies in the law of gravity. Below this size, if theory is correct, gravitation would no longer obey an inverse-squared type law. Instead, it would drop off at a different rate. Experimentalists are galvanized by the possibility of detecting such an effect.

In 2001, the Eöt-Wash research group announced the results of an experiment designed to measure the submillimeter behavior of gravity. The Eöt-Wash group is a team of University of Washington researchers led by Eric Adelberger. It gets its name from a clever play on words: "Eöt-Wash" is pronounced like "Eötvös," the name of a famous Hungarian gravitational physicist.

The team tested the law of gravity with an extremely precise device they designed. It consisted of a ring-shaped aluminum pendulum delicately suspended just above a set of two slowly rotating copper disks, one thicker than the other. The disks were arranged such that if the inverse-squared law of gravity were correct, their effects on the pendulum would exactly cancel out. In contrast, if gravity on the submillimeter scale obeyed a different relationship, the pendulum would experience a net torque (rotational tendency) and begin to twist slightly.

At first the researchers seemed to discover a slight pull on the ring, appearing to violate the hallowed gravitational law. As hard as they tried, they couldn't explain this motion. Finally, one of the team members figured out its cause: the commercially produced fiber suspending the pendulum was labelled wrong. It could move only 98 percent as far in one of the directions as it was supposed to. Once they took this into account, the violation completely vanished. The inverse-squared law of gravity appeared correct down to scales at least as small as $\frac{1}{5}$ of one millimeter (less than $\frac{1}{100}$ of one inch). The researchers have subsequently made several improvements to extend their findings to an even tinier scope.

The Eöt-Wash results came as something of a disappointment to advocates of large extra dimensions. The team's precise vindication of Newton's relationship appeared to rule out some of the simplest approaches. However, as Dimopoulos has remarked, it is "too early to tell" if his collaborative model is still viable. "Experiments at smaller length scales are coming up," he eagerly points out.[13]

Another group of dimension hunters, led by John Price and Joshua Long of the University of Colorado, have employed a different apparatus to record gravitational discrepancies. Instead of using rotating disks, they have based their experiment on thin tungsten reeds vibrating 1,000 times per second. Sensitive electronics have indicated that the standard law of gravity holds down to less than $\frac{1}{10}$ of a millimeter, barely twice the thickness of a strand of hair.

Other researchers have attempted to measure differences between the speeds of light and gravity. If found, such a distinction would offer some indication that gravitons are taking shortcuts through the bulk. Such proof doesn't seem soon to be forthcoming, however. In 2003, Sergei Kopeikin of the University of Missouri and Ed Fomalont of the National Radio Astronomy Observatory in Virginia announced the results of an astronomical study that found no significant difference between the two speeds.

Yet another technique for measuring extra dimensions requires only that they are compact. (In this context, "compact" means closed and bounded like a loop, not necessarily small.) It applies the Heisenberg uncertainty principle to particles produced in accelerators. According to that precept, close confinement in physical space leads to a large range of possibilities in momentum space. The same is true of particles restricted to the closed loops of extra dimensions. The outcome is a spectrum of highly energetic particles echoing the original, like voices reverberating through a cave. Following Einstein's dictum that energy is related to mass, these so-called Kaluza-Klein modes would be much heavier than the originals. They would present themselves in a "tower of states": a step-ladder of particles, one more massive than the next. Researchers hope to discover their telltale signatures in accelerator data, unveiling the ghostly influence of a higher-dimensional world.

Two of the most prominent experts in the search for higher dimensions are theorist Joseph Lykken and experimentalist Maria Spiropulu, both at the University of Chicago. They believe that high-energy gravitons escaping from our brane into the bulk could well make their presence known in certain types of particle collisions. As Spiropulu has told conference audiences, "Gravitons are the most robust probe of extra dimensions."[14]

The disappearing gravitons could not be seen directly before they slipped into another realm. Rather, their signature would be a jet of hadrons emanating from a gluon (boson carrying the strong force). These hadrons would emerge in such a way that they would tell the story of gravitons gone missing. Detector readings would provide valuable clues about the lost piece of the puzzle.

In 2001, a group of researchers led by Greg Landsberg of Brown University reexamined sixty million particle collisions that had taken place at the Tevatron during a previous four-year period. Digging through this mountain of archeological evidence, they sifted for signs of missing gravitons in any of the recorded interactions. Alas, they found no lost gold.

Naturally, once the LHC goes on line, experimentalists will be lining up in droves to test various higher-dimensional theories. Fervently searching for Kaluza-Klein modes of existing particles, they'll also look for credible signs of absentee gravitons. All the while, they'll scan the data for supersymmetric companions and keep their eye out for that long-sought trophy, the Higgs boson. With all the potential wonders to reveal, it will truly be a glorious age for experimental physics.

Conclusion

Extra-dimensional Perception

The physics of extra dimensions is a revolution in the making. If the extra dimensions exist, the visible universe is just a slice. The rest is terra incognita.

—Joseph Lykken, lecture at a meeting of the American
Physical Society

The Dimensionality Revolution

At a meeting of the American Physical Society in 2003, Joseph Lykken gave a talk about the experimental prospects for Kaluza-Klein theory and brane world models, entitled "Mysteries of Extra Dimensions." Lykken's opening statement was a nod to traditional undercurrents of hesitancy about the subject: "It looks like from the title of this talk that this is some kind of psychic phenomena, but I hope to convince you it's a physics talk."[1]

Indeed, experimentalists took a lot of convincing before they were willing to wage their careers on trying to find hidden extra-dimensional portals. Years ago one would be hard pressed to find reference to higher dimensions in any experimental papers. Those who spent their days rigging up detectors and sorting through collision data would barely conceal their mirth when listening to theorists' notions of ten-, eleven-, or twenty-six-dimensional universes. Chuckling to themselves, they might wonder if such far-flung research even deserved to be called physics, rather than simply abstract philosophy.

In those less-enlightened times, submitting articles on Kaluza-Klein theory to a major journal could be like Russian roulette. One could easily end up with a reviewer who thinks that the subject is only fit for crackpots. Alan Chodos remembers receiving a referee report stating in essence, "There's no basis for any of this extra-dimensional stuff. Until it's proven this paper should not be published."[2]

Feelings about higher dimensions were such that most of the theorists who worked on the topic were plagued by second thoughts. Though excited about it at first, many decided to veer more toward the mainstream. That was certainly true for Klein, who abandoned five-dimensional theory for a time to concentrate on quantization, and for Kaluza, who eventually put his model aside to focus primarily on teaching, textbook writing, and his family. Though hardly conventional thinkers, Einstein and Pauli also vacillated in their attitudes, sometimes even from year to year. Einstein's assistants Bergmann and Bargmann simply put the topic aside once they went off on their own. Changes of heart seemed to pervade this controversial approach.

Chodos recalls how his collaborator, Thomas Appelquist, eventually decided to close the box on higher dimensions: "At some point he disengaged from that and I kept going. He said, 'enough already. I don't want to be known as the guy who does extra dimensions. I want to do more nuts and bolts physics.' So even though he wrote some papers and co-authored the book, he would be on the side of not believing this stuff."[3]

Given this chronicle of ambivalence, it is a tribute to contemporary string theorists that they've managed to persuade the experimental community to give their ideas a shot. They have performed this feat in part by pointing out all the issues that extra dimensions could possibly resolve. These include the hierarchy problem of why gravity is so much weaker than its brethren forces, the related cosmological constant and dark energy dilemmas of why the universe is accelerating at a particular rate, the puzzle of how to quantize gravity, and finally the question of finding a unified vision of nature. Moreover, thanks to Maldacena's conjecture, extra-dimensional models may help to unravel certain difficulties in quantum chromodynamics and other standard gauge theories. The potential for addressing all these long-standing enigmas offers a powerful case indeed for taking M-theory and extra dimensions seriously.

None of these arguments would justify lab space and beam time if there weren't ways of designing experiments suitable to test the extra-dimensional hypothesis. Fortunately there are—as an exciting array of research groups have demonstrated—ranging from tabletop designs to projects for gargantuan accelerators. The Large Hadron Collider, in particular, offers great promise for resolving some of these questions. Who knows what secrets it will reveal when it becomes fully operational in the late 2000s?

Reading, Writing, and Hyperspace

What if it's all true? What if tests show conclusively that Kaluza, Klein, and their successors were right about the hidden recesses of space? Suppose higher dimensions prove essential for describing all aspects of physics with a simple set of principles. How would this momentous discovery change our culture? In particular, how would science convey these ideas to a public used to associating higher dimensions with the occult, if it considers them at all?

It took generations for scientists to persuade the populace that Earth revolves around the Sun. A century and a half after Darwin's proposal, segments of the public remain unconvinced. One would hope that a final theory of the universe could be appreciated by all, perhaps even taught on some level in school. Yet because our senses are limited, explaining this topic in a tangible way would present a critical pedagogical challenge.

One might turn for guidance to the lessons of the late nineteenth century, when Riemann, Cayley, and others first brought higher dimensions to mathematics. Gradually the idea of hyperspace seeped into the popular imagination through geometric models, lectures, essays, and stories. Competing with pseudoscientific interpretations such as Zöllner's and the theosophists', certain rudimentary notions about higher dimensions acquired a measure of respect—at least among scientifically literate audiences. Eventually these novel conceptions found expression in the world of art—in futurism and cubism, for example—bolstering general awareness even further.

In many ways a ten- or eleven-dimensional unified model such as M-theory represents even more of a challenge. It's hard enough to envision hypercubes, let alone Calabi-Yau shapes or dualities between branes and strings. If such elements assume pivotal roles in the scientific canon, it would take extraordinary teaching tools to enable students to fathom their structures. Reportedly Witten can manipulate such constructs in his head while gazing out his office window, but what about the rest of us mere mortals?

Perhaps science classes of the future would come equipped with virtual reality simulators, enabling immersion into hyperworlds. Users would manipulate projections of higher-dimensional objects, learning to understand how they fit together. In such a manner, maybe "extra-dimensional perception" could be taught to even young children, coaching their developing brains to encode complex geometric relationships. Like learning a foreign language, a new generation of thinkers would come to be fluent in hyperspace.

Sketches of the Impossible

Fortunately, a number of visionary computer scientists and artists have begun to lay the groundwork for hyperspace education. Using enhanced graphics, animation, stereoscopy, and other tools, they have fashioned ways for the eye to see the seemingly impossible. Like high-tech Hintons, their goal is to bring higher dimensions into the realm of experience—much like slow-motion photography has made many aspects of nature much more vivid.

One of the leaders in this movement is Thomas Banchoff, professor of mathematics at Brown University. Working with colleagues at Brown's Computer Graphics Laboratory, he has pioneered the use of motion, projection, and perspective to bring higher-dimension shapes to life. Numerous students have worked with him in learning how to render the bizarre configurations of hyperspace objects as they rotate through various angles and are intersected by different planes. Some of his award-winning films include *The Hypercube: Projection and Slicing*, which he produced on the computer with Charles

Strauss. Banchoff is also a recognized scholar on the life and work of Edwin Abbott.

An artist greatly influenced by Banchoff is New York–based painter, sculptor, and architect Tony Robbin. The founder of Pattern Painting, Robbin finds delightful expression in filling canvases with colorful geometric designs. He created his earliest pieces with a spray gun and stencils, setting down layers of texture like patterned quilts.

Then, Robbin encountered Banchoff's multidimensional imagery—a momentous occasion. He knew then and there that he wanted to explore higher dimensions in his artistic creations. As he recalls, "For three nights, I woke frequently from dreams of the images I had seen on Banchoff's computer: the green screen, the quivering geometric figure. It seemed as if these images were imprinted on my mind. I had seen the fourth dimension directly. Here at last was the secret I needed to fulfill my ambitions."[4]

Training himself to be an expert programmer, Robbin developed some of the most sophisticated hyperspace graphical routines. He used these to arrange three-dimensional figures into designs suggestive of a higher-dimensional unity. These pieces can be appreciated on their own, or with the help of 3-D glasses.

Another contemporary artist, filmmaker Peter Rose, posits a spatialized form of time as the fourth dimension. In films such as *The Man Who Could Not See Far Enough,* he suggests ways of transcending our sensory limitations and assuming a perspective beyond immediate experience. Using time delays and other techniques, Rose weaves together past, present, and future into kaleidoscopic images, turning the temporal axis on its side and rendering it spacelike.

Rose first learned about higher dimensions from Gamow's book *One Two Three . . . Infinity,* as well as from a number of well-known science fiction stories. He also remembers a psychedelic experience in which he saw "thousands of time-delayed images in the interstices of a napkin." This inspired some of his first multi-image effects.

Rose sees a great potential for conveying higher-dimensional notions to the public. "Computer animation has certainly arrived at the point where simulations of complex hyper-dimensional behavior can be convincingly suggested," he says, "but it is in popular culture that one finds the most direct address. *Star Trek,* among a large number of other films, has grappled with questions about time, causality, space, etc. I would conjecture that many people are comfortable, in general terms, with concepts that would have been inaccessible several decades ago."[5]

Indeed, as in the days of Heinlein and Deutsch, science fiction continues to be a construction kit with which extraordinary ideas about the universe can be assembled. Today, two of the most imaginative writers in that genre, Rudy Rucker and Ian Stewart, are university mathematics professors who, in the tradition of Abbott and Carroll, have sought whimsical ways of conveying their ideas.

Rucker is a man of many hats. Some recognize his scholarship on the history of the fourth dimension, particularly his texts on the subject and his edited volume of Hinton's works. Others regard him as an important contributor to the science of complexity, particularly the discrete, self-organizing world known as cellular automata. Yet others know him as one of the founders of the cyberpunk movement, a late twentieth-century genre that emphasized freewheeling lifestyles, bizarre hairstyles, and virtuosity with computer files.

Within the context of this movement and another that he calls "transrealism," Rucker has drawn a loyal following as a weaver of fabulous tales, many of which are set in alternate universes or hyperspace. Denizens of the "Ruckerverse" include Babs, a sexy hypersphere that exerts an irresistable allure over men, and Joe Cube, a three-dimensional hero who, in the A Square tradition, is wrested into a higher dimension.

Impressed by Banchoff and Robbin's creations, Rucker has offered his research students their own multidimensional simulations to complete. He strongly believes in the mind's ability to fathom worlds beyond space. "The brain is a fabulously complex web of messages and connections," says Rucker. "Even though the matter of the brain is strictly three-dimensional there is no reason the brain cannot build up an accurate model or 'dynamic shadow' of a four-dimensional construct."[6]

Another writer/mathematician who has offered both scholarly and fictional treatments of hyperspace is University of Warwick professor Ian Stewart. Stewart, a regular columnist for *Scientific American* and a frequent contributor to *New Scientist,* has just published an annotated version of *Flatland.* His sequel to that book, called *Flatterland,* extends the chronicle to include a humorous look at contemporary multidimensional theories.

A number of other writers have successfully conveyed to popular audiences the subtleties of modern string theory and hyperspace worlds. Recent works by scientists Brian Greene, Michio Kaku, and Clifford Pickover have offered lucid introductions to the subject. Even Michael Duff has joined the fray, penning a clever parody of *Flatland* involving ten-dimensional string theorists coming to terms with eleven-dimensional M-theory.[7] If the universe does turn out to be a kaleidoscope of extra dimensions, lay readers trying to grasp a "Theory of Everything" will undoubtedly be in good hands.

The Technology of Tomorrow

There is something depressing about the phrase "final theory," as if once it's found all science will grind to a halt. Although a unified model of all known forces would be a stunning achievement, it certainly wouldn't represent the end of human inquiry. At the very least, centuries of interpretation and applications would more than justify physicists' salaries.

As one possible avenue of research—if we happen to live in a multidimensional cosmos—scientists could attempt to find short cuts through the fabric of space. These would allow rapid communication or even passage between one part of the universe and another. The transmission of information and/or materials at faster-than-light speeds would offer untold benefits for the exploration of the cosmos, circumventing the vast distances normally involved.

Geometric connections between separate regions of space were first proposed by Einstein and Rosen in 1935. Known as Einstein-Rosen bridges or wormholes, they join two flat manifolds together with a curved throat. Einstein, Rosen, and later Wheeler tried to use these to fashion particle theories. Subsequently, other researchers such as Kip Thorne of Caltech have speculated that specially designed wormholes could be used for interstellar travel. If they could be rendered large, safe, and stable, they could hypothetically act as subways between remote parts of the universe.

According to the Manyfold Universe scheme, connections would exist of a different sort. While the material components and electromagnetic radiation of our world would be trapped in a kind of Flatland, gravitons could leak out to other folds. Thus gravity alone would monopolize all the short cuts.

Although we couldn't directly use such passages ourselves, we could conceivably harness gravitons to transfer energy and information through them. Passing from one fold of our brane to another, these particles could circumvent lengthy journeys through physical space. Thus they could effectively provide faster-than-light communication.

We would first need to develop reliable transmitters and detectors of gravitational waves. To send messages, the transmitters would modulate gravitational signals, which would freely pass through the bulk. Then distant detectors would convert the waves back into readable information content. Using such a system, for example, we could communicate with extremely remote spaceships, as long as their distance across the bulk is not too great.

Naturally, given the current state of technology, such a mechanism would be a long way off. Moreover, it would depend upon particular types of brane world models turning out to be correct. However, it's fun to dream about possible future applications for unified theories of physics. At the very least, they help dispel worries that a "final theory" would mark the end of scientific inquiry.

The Legacy of Kaluza and Klein

It's been a long, strange journey from the bridges of Königsberg to the tunnels of the Large Hadron Collider, and from the streets of Ann Arbor to the byways of escaping gravitons. If our path has been circuitous, it is only because we have followed the twists and turns of extra dimensions. If some of the theorists we've met seem to have a lot on their minds, imagine trying to wrap your brain around hyperspace as part of your everyday research.

What drives a thinker to set aside familiar spatial boundaries and contemplate the great beyond? Why consider bizarre scenarios that bear scarce resemblance to our sensory experiences? Given all the opportunities in ordinary physics, why search for something extraordinary?

Perhaps it is the human aversion to limits. We want to know what is just outside the frontiers of knowledge. It disturbs us to be told, "No trespassing beyond this point." If nature counts to three, we want to count to four, five, or more.

For Kaluza and Klein, it was a desire for completeness. The fifth dimension offered the extra quarters needed to accommodate all of nature's lodgers. Why should electromagnetism be left out in the cold, while gravity rests in its own cozy space-time? And, as Klein added, why should quantum theory live in a fuzzy framework, when it could also reside in something more solid?

Today theorists are trying to house an even greater number of disparate ideas. They are hoping to build an edifice that will sustain a successful pairing between electromagnetism and the weak force, an elegant representation of the strong interaction, and a geometric model of gravity—as well as quantum theory itself—reflecting each of these forces' strengths and weaknesses in the building design. It would also require a vast substructure to contain dark matter, dark energy, and other mysterious elements. Will this towering castle reach high into the clouds of ethereal new dimensions? Only time will reveal the hidden architecture of the cosmos.

Notes

Introduction: The Kaluza-Klein Miracle

1. Gary Gibbons, "Brane-Worlds," *Science* 287 (January 7, 2000): 49.
2. Interview with Stanley Deser, Brandeis University, November 22, 2002.

1   The Power of Geometry

1. H. G. Wells, "The Time Machine," in *Seven Science Fiction Novels of H. G. Wells* (New York: Dover, 1934), p. 4.
2. Diderot and Jean D'Alembert, *Encyclopédie ou Dictionnaire Raisonné des Sciences des Arts et des Métiers,* vol. 4 (Stuttgart: Friedrich Frommann, 1988), p. 1010.
3. R. C. Archibald, "Time as a Fourth Dimension," *Bulletin of the American Mathematical Society* (May 1914): 409–12.
4. Immanuel Kant, "Thoughts on the True Estimation of Living Forces," in *Kant's Inaugural Dissertation and Early Writings on Space,* trans. John Handyside (Chicago: Open Court, 1929), p. 13.

2   Visions of Hyperspace

1. Eric Temple Bell, *Men of Mathematics* (New York: Simon and Schuster, 1937), p. 221.
2. Ibid., p. 220.
3. Leonard Mlodinow, *Euclid's Window: The Story of Geometry from Parallel Lines to Hyperspace* (New York: Simon and Schuster, 2001), p. 119.
4. Bell, *Men of Mathematics,* p. 497.
5. David Wells, *The Penguin Book of Curious and Interesting Mathematics* (New York: Penguin, 1997), p. 139.
6. Letter from J. J. Sylvester to Joseph Henry, April 12, 1846. Reprinted in Karen Hunger Parshall, *James Joseph Sylvester: Life and Work in Letters* (Oxford: Clarendon Press, 1998), p. 15.

7. James J. Sylvester, "A Plea for the Mathematician," *Nature* 1 (December 30, 1869): 238.

8. Gauss's use of a bookworm analogy is described in his 1856 biography by Sartorius von Waltershausen. In an 1870 talk, the mathematician Hermann von Helmholtz similarly refered to two-dimensional beings living on a spherical surface.

9. Sylvester, "A Plea for the Mathematician," p. 238.

10. Ibid.

11. Claude Bragdon, *Four-Dimensional Vistas* (New York: Alfred A. Knopf, 1916), p. 38.

12. Helena P. Blavatsky, *The Secret Doctrine*, vol. 1 (London: Theosophical Publishing Society, 1888), pp. 251–52.

13. C. W. Leadbeater, quoted in Alexander Horne, *Theosophy and the Fourth Dimension* (London: Theosophical Publishing Society, 1928), p. vii.

14. Charles Howard Hinton, *The Fourth Dimension* (London: Allen and Unwin, 1904), p. 207.

15. John Stegall, quoted in Thomas Hinde, *Carpenter's Children: The Story of the City of London School* (London: James and James, 1994), p. 66.

16. Edwin Abbott Abbott, *Flatland: A Romance of Many Dimensions* (London: Seeley and Co., 1884), p. ii.

17. Rudy Rucker, *The Fourth Dimension* (Boston: Houghton Mifflin, 1984), p. 11.

18. Abbott, *Flatland*, p. ii.

19. Ibid.

20. James E. Beichler, "The Psi-ence Fiction of H. G. Wells," *YGGDRASIL: The Journal of Paraphysics,* http://ourworld.compuserve.com/homepages/Paraphys/psifi.htm, pp. 3–4.

21. S. (pseudonymous author), "Four-Dimensional Space," *Nature* 31 (March 26, 1885): 481.

22. Bernard Bergonzi, *The Early H. G. Wells: A Study of the Scientific Romances* (Manchester, England: Manchester University Press, 1969), p. 3.

23. H. G. Wells, "The Plattner Story," in *28 Science Fiction Stories of H. G. Wells,* (New York: Dover, 1952), p. 444.

24. Wells, "The Time Machine," p. 5.

25. Ibid., p. 4.

26. Simon Newcomb, "Modern Mathematical Thought," *Nature* 49 (February 1, 1894): 328–29.

3   The Physicist's Stone: Uniting Electricity, Magnetism, and Light

1. James Clerk Maxwell, "Lines written under the conviction that it is not wise to read mathematics in November after one's fire is out," reprinted in Lewis Campbell and William Garnett, *The Life of James Clerk Maxwell* (London: Macmillan and Co., 1882), pp. 622–25.

2. James Clerk Maxwell, "Report on Tait's lecture on force," reprinted in Campbell and Garnett, *The Life of James Clerk Maxwell,* p. 647.

3. James Clerk Maxwell, "A Paradoxical Ode," reprinted in Campbell and Garnett, *The Life of James Clerk Maxwell,* pp. 649–50.

4. Albert Einstein, *Autobiographical Notes,* trans. and ed. Paul Arthur Schilpp, (La Salle, Ill.: Open Court, 1979), p. 49.

5. Peter Galison, "Minkowski's Space-Time: From Visual Thinking to the Absolute World," in *Historical Studies in the Physical Sciences,* vol. 10, ed. Russell McCormmach, Lewis Pyenson, and Roy Steven Turner (Baltimore: Johns Hopkins, 1979).

6. Henri Poincaré, "Science and Hypothesis," in *The Value of Science: Essential Writings of Henri Poincaré,* ed. Stephen Jan Gould (New York: Modern Library, 2001), p. 48.

7. Henri Poincaré, "Science and Method," in *The Value of Science,* p. 438.

8. Hermann Minkowski, address delivered at the 80th Assembly of German Natural Scientists and Physicians, September 21, 1908.

9. Albert Einstein's Third Year Curriculum Performance Report, 1898–1899, in *The Collected Papers of Albert Einstein,* vol. 1, John Stachel, ed., and Anna Beck, trans. (Princeton, N.J.: Princeton University Press, 1995), p. 27.

10. Barry Parker, *Einstein's Dream: The Search for a Unified Theory of the Universe* (New York: Plenum, 1986), p. 31.

11. Einstein, *Autobiographical Notes,* p. 15.

12. Abraham Pais, *Subtle Is the Lord: The Science and the Life of Albert Einstein* (New York: Oxford University Press, 1982), p. 152.

13. Dennis Overbye, *Einstein in Love* (New York: Viking, 2000), p. 160.

14. Albert Einstein and Jakob Laub, "Über die elektromagnetischen Grundgleichungen für bewegte Körper," *Annalen der Physik* 26 (1908): 532; translated and quoted in John Stachel, *Einstein from "B" to "Z"* (Boston: Birkhäuser, 2002), p. 287.

15. Overbye, *Einstein in Love,* p. 160.

16. Letter from Albert Einstein to Arnold Sommerfeld, July 1910, in *The Collected Papers of Albert Einstein,* vol. 1, Robert Schulmann, ed., and Anna Beck, trans. (Princeton, N.J.: Princeton University Press, 1995), p. 157.

17. Albert Einstein, "The Theory of Relativity," lecture given at the meeting of the Zurich Naturforschende Gesellschaft, January 16, 1911, in *The Collected Papers of Albert Einstein,* vol. 3, A. J. Kox, Martin J. Klein, and Robert Schulmann, eds., and Anna Beck, trans. (Princeton, N.J.: Princeton University Press, 1993), p. 350.

18. Letter from Einstein to A. Kleiner, April 3, 1912.

19. Albert Einstein and Leopold Infeld, *The Evolution of Physics* (New York: Simon and Schuster, 1938), p. 207.

20. Interview with Gerald Holton, Harvard University, November 22, 2002.

21. Paul M. Laporte, "Cubism and Relativity (With a Letter of Albert Einstein), *Art Journal* 25 (1966): 246.

4   *Getting Gravity in Shape*

1. Stachel, *Einstein from "B" to "Z"* (Boston: Birkhäuser, 2002), p. 262.
2. Louis Kollross, "Albert Einstein en Suisse-Souvenirs," *Fünfzig Jahre Relativitätstheorie* (Bern, 11–16 July 1955), *Helvetica Physica Acta, Supplementum IV* (1956): 274–75.
3. Letter from Albert Einstein to C. Seelig, May 5, 1952, quoted in Albrecht Fölsing, *Albert Einstein*, Ewald Osers, trans. (New York: Penguin, 1997), p. 296.
4. Albert Einstein and A. D. Fokker, "Die Nordströmsche Gravitationstheorie vom Standpunkt des absoluten Differentialkalküls," *Annalen der Physik* IV Folge, 44 (1914).
5. Letter from Albert Einstein to Michele Besso, December 10, 1915, in *Albert Einstein-Michele Besso, Correspondance*, Pierre Speziali, ed. (Paris: Hermann, 1972), p. 59.
6. Interview with John Wheeler, Princeton University, November 5, 2002.
7. Eva Isaksson, "Gunnar Nordström: On Gravitation and Relativity" (paper presented at the XVIIth International Congress of the History of Science, University of California, Berkeley, August 1985).
8. Thomas Appelquist, Alan Chodos, and Peter Freund, *Modern Kaluza-Klein Theories* (Reading, Mass.: Addison-Wesley, 1987), p. 10.
9. Gunnar Nordström, "On the Possibility of a Unification of the Electromagnetic and Gravitational Fields," *Phys. Zeitsch.* 15 (1914): 504, trans. Peter Freund; in Appelquist, Chodos, and Freund, *Modern Kaluza-Klein Theories*, p. 52.
10. Ibid.
11. Paul Ehrenfest, Research Notebook, Museum Boerhaeve, Leiden.
12. Martin J. Klein, *Paul Ehrenfest: The Making of a Theoretical Physicist* (London: North-Holland, 1970), p. 309.
13. Ibid.
14. Interview with John Stachel, Boston University, July 30, 2002.
15. Paul Ehrenfest, "In what way does it become manifest in the fundamental laws of physics that space has three dimensions?" *Proceedings of the Royal Academy* (Amsterdam), 20 (1917): 200–209.
16. Phone interview, Martin J. Klein, September 11, 2002.
17. Letter from Hermann Weyl to Albert Einstein, March 1, 1918, in *The Collected Papers of Albert Einstein*, vol. 8, Robert Schulmann and A. J. Kox, eds., and Ann Hentschel, trans. (Princeton, N.J.: Princeton University Press, 1995).
18. Letter from Albert Einstein to Hermann Weyl, March 8, 1918, in ibid.
19. Ibid.

20. Letter from Albert Einstein to Hermann Weyl, April 8, 1918, in ibid., p. 522.
21. Letter from Hermann Weyl to Albert Einstein, May 19, 1918, in ibid., p. 562.
22. Letter from Albert Einstein to Hermann Weyl, November 29, 1918, in ibid., p. 699.
23. Letter from Hermann Weyl to Albert Einstein, December 20, 1918, in ibid., p. 709.

5 Striking the Fifth Chord: Kaluza's Remarkable Discovery

1. Daniela Wünsch, private communication, July 16, 2002.
2. Daniela Wünsch, "Theodor Kaluza: Leben und Werk (Life and Work)," vol. 1 (Ph.D. diss., University of Stuttgart, 2000), p. 127.
3. Ibid., p. 24.
4. Ibid., p. 11.
5. Varadaraja V. Raman, "Theodor Kaluza," in Charles Gillespie, ed., *Dictionary of Scientific Biography* (New York: Scribner, 1970), p. 212.
6. Theodor Kaluza, Jr., *Erinnerungen (Remembrances)*, University of Hannover Mathematics Department reprint (1990), p. 24.
7. Wünsch, "Theodor Kaluza," p. 70.
8. Ibid., p. 66.
9. Kaluza, *Erinnerungen*, p. 24.
10. Interview with Martin Kneser, Göttingen, January 8, 2003.
11. Varadaraja V. Raman, "Theodor Kaluza," in Gillespie, *Dictionary of Scientific Biography*, p. 212.
12. Kaluza, *Erinnerungen*, p. 24.
13. Ibid.
14. Daniela Wünsch, "Einstein, Kaluza and the Fifth Dimension" (talk presented at the Sixth Conference on the History of General Relativity, Amsterdam, July 26–29, 2002).
15. Theodor Kaluza, Jr., interview on "NOVA: What Einstein Never Knew," originally broadcast October 22, 1985.
16. Theodor Kaluza, "On the Unity Problem of Physics," trans. Peter Freund, T. Muta, and C. Hoensalaers, reprinted in Thomas Appelquist, Alan Chodos, and Peter Freund, *Modern Kaluza-Klein Theories* (Reading, Mass.: Addison-Wesley, 1987), p. 61.
17. Ibid., p. 62.
18. Letter from Albert Einstein to Theodor Kaluza, April 21, 1919, trans. C. Hoensalaers, in Venzo De Sabbata and Ernst Schmutzer, eds., *Unified Field Theories of More than four Dimensions* (Singapore: World Scientific, 1983), p. 449.
19. Letter from Albert Einstein to Theodor Kaluza, April 28, 1919, in ibid., p. 451.

20. Letter from Albert Einstein to Theodor Kaluza, October 14, 1921, in ibid., p. 451.
21. Kaluza, "On the Unity Problem of Physics," p. 68.
22. Letter from Albert Einstein to Hermann Weyl, June 6, 1922 (Einstein Archives).
23. Letter from Albert Einstein to Paul Ehrenfest, August 18, 1925 (Einstein Archives).
24. Letter from Theodor Kaluza to Albert Einstein, February 6, 1925 (Einstein Archives).
25. Letter from Albert Einstein to Theodor Kaluza, February 27, 1925, in *Unified Field Theories,* p. 457.
26. Letter from Albert Einstein to Karl Herzfeld, probably January 1927 (Einstein Archives).
27. Postcard from Albert Einstein to Paul Ehrenfest, September 3, 1926 (Einstein Archives).

6   Klein's Quantum Odyssey

1. Interview with Stanley Deser, Brandeis University, November 22, 2002.
2. Oskar Klein, "From My Life of Physics," reprinted in *Oskar Klein Memorial Lectures,* vol. 1 (Singapore: World Scientific, 1989), p. 106.
3. Interview with Stanley Deser, Brandeis University, November 22, 2002.
4. Niels Bohr, quoted by Edward Teller in S. Rozental, ed., *Niels Bohr* (New York: Wiley, 1967), p. 103.
5. Klein, "From My Life of Physics," p. 107.
6. Oskar Klein, interview with John L. Heilbron and Léon Rosenfeld, Copenhagen, February 20, 1963.
7. John Stachel, "Einstein and 'Zweistein,' " in John Stachel, *Einstein from "B" to "Z"* (Boston: Birkhäuser, 2002), p. 540.
8. Oskar Klein, *Kosmos* 37 (1959): 9.
9. Abraham Pais, "Glimpses of Oskar Klein as Scientist and Thinker," in *Proceedings of the Oskar Klein Centenary Symposium,* Ulf Lindström, ed. (Singapore: World Scientific, 1995), p. 8.
10. Klein, interview with Heilbron and Rosenfeld.
11. Klein, "From My Life of Physics," p. 108.
12. Ibid., p. 111.
13. Klein, interview with Heilbron and Rosenfeld.
14. Abraham Pais, *Subtle Is the Lord: The Science and the Life of Albert Einstein* (New York: Oxford University Press, 1982), p. 332.
15. George Uhlenbeck, interview with Thomas S. Kuhn, Rockefeller Institute, April 5, 1962.
16. Letter from Paul Ehrenfest to Albert Einstein, August 26, 1926 (Einstein Archives).
17. Werner Heisenberg, in Rozental, *Niels Bohr,* p. 103.

18. Letter from Albert Einstein to Max Born, December 4, 1926, in *Albert Einstein-Max Born, Briefwechsel (Correspondence)*, Max Born, ed. (Munich, 1969), p. 129.

19. Letter from Oskar Klein to Paul Ehrenfest, March 1930, Archives for the History of Quantum Physics.

20. Letter from Paul Ehrenfest to Oskar Klein, March 19, 1930.

21. Oskar Klein, "On Political Quantization," unpublished letter submitted to the *Journal of Jocular Physics* (1935; Niels Bohr Archives).

22. Oskar Klein, interview with Thomas S. Kuhn and John L. Heilbron, Copenhagen, July 16, 1963.

23. Ibid.

24. Ibid.

25. Letter from Heinrich Mandel to Oskar Klein, December 4, 1928 (Niels Bohr Archives).

26. Abraham Pais, "Glimpses of Oskar Klein as Scientist and Thinker," in Lindström, *Proceedings of the Oskar Klein Centenary Symposium*, p. 14.

7   Einstein's Dilemma

1. Albrecht Fölsing, *Albert Einstein*, Ewald Osers, trans. (New York: Penguin, 1997), p. 637.

2. Letter from Albert Einstein to the Lebach family, January 16, 1931, ibid., p. 636.

3. Charles Chaplin, *My Autobiography* (New York: Simon and Schuster, 1964), p. 320.

4. Ibid., p. 252.

5. Ibid., p. 321.

6. Ibid.

7. Albert Einstein, "Isaac Newton," *Nature* 119 (1927): 467; *Science* 65 (1927): 347.

8. Fölsing, *Albert Einstein*, p. 637.

9. Letter from Albert Einstein to Niels Bohr, May 2, 1920, in Banesh Hoffman with Helen Dukas, *Albert Einstein: Creator and Rebel* (New York: Viking, 1972), p. 178.

10. Ibid.

11. Postcard from Albert Einstein to Michele Besso, May 1, 1926, in *Albert Einstein-Michele Besso, Correspondance*, Pierre Speziali, ed. (Paris: Hermann, 1972).

12. Letter from Albert Einstein to Vero and Bice Besso, March 21, 1955, in Fölsing, *Albert Einstein*, p. 741.

13. Letter from Albert Einstein to Paul Ehrenfest, January 11, 1927 (Einstein Archives).

14. Ruth Moore, *Niels Bohr: The Man and the Scientist* (London: Hodder and Stoughten, 1967), p. 164.

15. W. Heisenberg, in *Niels Bohr: A Centenary Volume,* A. P. French and P. J. Kennedy, eds. (Cambridge, Mass.: Harvard University Press, 1985), pp. 170–71.
16. Letter from Albert Einstein to Paul Ehrenfest, January 21, 1928 (Einstein Archives).
17. "Einstein Extends Relativity Theory," *New York Times,* January 11, 1929, p. 1.
18. "Einstein Distracted by Public Curiosity; Seeks Hiding Place," *New York Times,* February 4, 1929, p. 1.
19. Letter from Richard Tolman to Paul Ehrenfest, January 27, 1931 (Ehrenfest Archives).
20. Albert Einstein and Leopold Infeld, *The Evolution of Physics* (New York: Touchstone, 1967), p. 310.
21. Albert Einstein and Walther Mayer, "Einheitliche Theorie von Gravitation und Elektrizität [Unified Field Theory of Gravitation and Electricity]," *Sitzungsberichte der Preussischen Akademie der Wissenschaften zu Berlin* (1931), p. 541; in Fölsing, *Albert Einstein,* p. 648.
22. Wolfgang Pauli, review of *Ergebnisse der exakten Naturwissenschaften,* 10, Band, *die Naturwissenschaften 20,* pp. 186–87, in John Stachel, *Einstein from "B" to "Z"* (Boston: Birkhäuser, 2002), p. 544.
23. Letter from Albert Einstein to Wolfgang Pauli, January 22, 1932, in Abraham Pais, *Subtle Is the Lord: The Science and the Life of Albert Einstein* (New York: Oxford University Press, 1982), p. 347.
24. Fölsing, *Albert Einstein,* p. 649.
25. Letter from Albert Einstein to Abraham Flexner, July 30, 1932, in Pais, *Subtle Is the Lord,* p. 493.
26. Chaplin, *My Autobiography,* p. 322.
27. Interview with Gerald Holton, Harvard University, November 22, 2002.
28. Phone interview with Martin J. Klein, September 11, 2002.
29. Letter from Paul Ehrenfest to Richard Tolman, February 3, 1931 (Archives for the History of Quantum Physics).
30. Phone interview with Martin J. Klein, September 11, 2002.

8　Truth under Exile: Theorizing at Princeton

1. Letter from Oswald Veblen to Albert Einstein, October 28, 1931 (Einstein Archives).
2. Banesh Hoffmann, interview with Albert Tucker, "The Princeton Mathematics Community in the 1930s," October 13, 1984, transcript no. 20.
3. Letter from Banesh Hoffmann to Albert Einstein, February 11, 1932 (Einstein Archive).
4. Banesh Hoffmann, "Reminiscences," in *Albert Einstein: Historical and Cultural Perspectives,* Gerald Holton and Yehuda Elkana, eds. (Princeton, N.J.: Princeton University Press, 1982), p. 401.
5. Ernest Bergmann, personal communication, December 25, 2002.

6. Ibid.

7. Phone interview with Engelbert Schucking, December 9, 2002.

8. John R. Klauder, "Valentine Bargmann," *Biographical Memoirs* (Washington, D.C.: National Academy Press, 1999), p. 38.

9. Letter from Peter Bergmann to Albert Einstein, March 14, 1936 (Einstein Archives).

10. Letter from Max Bergmann to Albert Einstein, December 31, 1940 (Max Bergmann Archives).

11. Jamie Sayen, *Einstein in America* (New York: Crown, 1985), p. 147.

12. Interview with John Wheeler, Princeton University, November 5, 2002.

13. Banesh Hoffmann, "Working with Einstein," in *Some Strangeness in the Proportion: A Centennial Celebration to Honor the Achievements of Albert Einstein,* Harry Woolf, ed. (New York: Addison-Wesley, 1980), pp. 476–77.

14. Peter Bergmann, "Reminiscences," in Holton and Elkana, *Albert Einstein,* p. 398.

15. Phone interview with Peter Bergmann, June 27, 2002.

16. Valentine Bargmann, "Working with Einstein," in Woolf, *Some Strangeness in the Proportion,* p. 488.

17. Nathan Rosen, "Reminiscences," in ibid., p. 406.

18. Oskar Klein, interviewed with Thomas S. Kuhn and John L. Heilbron, Copenhagen, July 16, 1963.

19. Banesh Hoffmann, "Working with Einstein," p. 476.

20. Ibid., p. 477.

21. Peter Bergmann, *Introduction to the Theory of Relativity* (New York: Prentice-Hall, 1942), p. 272.

22. Albert Einstein and Peter Bergmann, "On a Generalization of Kaluza's Theory of Electricity," *Annals of Mathematics* 39 (1938): 683.

23. Letter from Wolfgang Pauli to Albert Einstein, September 6, 1938, in Wolfgang Pauli, *Wissenschaftlicher Briefwechsel (Scientific Correspondence),* vol. 2, 1930–1939, A. Hermann, K. v. Meyenn, and V. F. Weisskopf, eds. (Berlin: Springer, 1985), p. 598.

24. Letter from Albert Einstein to Wolfgang Pauli, September 19, 1938, in ibid., p. 601.

25. Jeroen van Dongen, "Einstein and the Kaluza-Klein Particle," *Studies in the History and Philosophy of Modern Physics* 33 (2002): 185.

26. Albert Einstein, Peter Bergmann, and Valentine Bargmann, "On the Five Dimensional Representation of Gravitation and Electricity," *Theodore von Kármán Anniversary Volume* (Pasadena: California Institute of Technology, 1941), pp. 224–25.

27. Letter from Léon Rosenfeld to Friedrich Herneck, 1962, in F. Herneck, *Einstein und sein Weltbild* (Berlin: Buchverlag der Morgen, 1976), p. 280.

28. Robert P. Crease and Charles C. Mann, *The Second Creation* (New York: Collier Books, 1986), p. 169.

29. Letter from Peter Bergmann to Albert Einstein, August 15, 1938 (Einstein Archives).

9   Brave New World: Seeking Unity in an Age of Conflict

1. Interview with Stanley Deser, Brandeis University, November 22, 2002.
2. Ibid.
3. Abdus Salam, "Nobel Prize Lecture," December 8, 1979.
4. David J. Gross, "Oskar Klein and Gauge Theory" (talk presented at the Oskar Klein Centenary Symposium, Stockholm, Sweden, 1994).
5. Interview with Stanley Deser, Brandeis University, November 22, 2002.
6. Letter from Albert Einstein to the president of the Bavarian Academy of Sciences, April 21, 1933, in Albert Einstein, *Ideas and Opinions,* Sonja Bargmann, trans. (New York: Bonanza Books, 1954), pp. 210–11.
7. Max Pinl and Lux Furtmüller, "Mathematicians Under Hitler," *Leo Baeck Yearbook,* Robert Weltsch, ed. (London: Secker and Warburg, 1973), p. 133.
8. Daniela Wünsch, "Theodor Kaluza: Leben und Werk (Life and Work), vol. 2 (PhD diss., University of Stuttgart, 2000), p. 57.
9. Ibid., pp. 59–61.
10. Ibid., pp. 72–74.
11. Theodor Kaluza and Georg Joos, *Höhere Mathematik für den Praktiker* (Leipzig: Barth, 1938).
12. Wünsch, *"Theodor Kaluza,"* p. 75.
13. Ibid., p. 86.
14. Interview with Gerald Holton, Harvard University, November 22, 2002.
15. Letter from Albert Einstein to Vern O. Knudson, June 1, 1940 (Einstein Archives).
16. Ernest Bergmann, personal communication, December 25, 2002.
17. Steven Weinberg, quoted in Dennis Overbye, "Obituary for Peter Bergmann," *New York Times,* October 23, 2002.
18. Lee Iacocca, *Iacocca: An Autobiography* (New York: Bantam, 1985), pp. 21–22.
19. Ernest Bergmann, personal communication, December 25, 2002.
20. Al McLennan, "History of the Physics Department," Lehigh University, http://www.physics.lehigh.edu/resources/history.pdf.
21. John R. Klauder, "Valentine Bargmann," *Biographical Memoirs* (Washington, D.C.: National Academy Press, 1999), p. 38.
22. Phone interview with Kenneth Ford, November 2002.
23. Engelbert Schucking, "Jordan, Pauli, Politics, Brecht, and a Variable Gravitational Constant," *Physics Today,* October 1999, p. 26.
24. Interview with Stanley Deser, Brandeis University, November 22, 2002.
25. Ibid.
26. Abraham Pais, *Subtle Is the Lord: The Science and the Life of Albert Einstein* (New York: Oxford University Press, 1982), p. 350.
27. Abraham Pais, "Wolfgang Ernst Pauli," *The Genius of Science* (New York: Oxford University Press, 2000), p. 216.

28. Freeman Dyson, "Another Visit with Wolfgang Pauli," *Physics Today,* August 2001, p. 11.

29. John Stachel, "Einstein and 'Zweistein,' " in *Einstein from "B" to "Z"* (Boston: Birkhäuser, 2002), p. 545.

30. Karl von Meyenn and Engelbert Schucking, "Wolfgang Pauli," *Physics Today,* (February 2001): 43.

31. Albert Einstein and Wolfgang Pauli, "On the Non-Existence of Regular Stationary Solutions of Relativistic Field Equations," *Annals of Mathematics* 44 (April 1943): 13.

32. Albert Einstein, *Autobiographical Notes,* Paul Arthur Schilpp, trans. and ed. (La Salle, Ill.: Open Court, 1949), p. 85.

33. Interview with John Wheeler, Princeton University, November 5, 2002.

34. Engelbert Schucking, "Jordan, Pauli, Politics, Brecht, and a Variable Gravitational Constant," *Physics Today,* October 1999, p. 27.

35. Ibid., p. 28.

36. Peter Bergmann, "Unified Field Theory with Fifteen Field Variables," *Annals of Mathematics* 49, no. 1 (January 1948): 255.

37. Letter from Wolfgang Pauli to Pascual Jordan, March 23, 1948, reprinted in Wolfgang Pauli, *Wissenschaftlicher Briefwechsel (Scientific Correspondence),* vol. 3, 1940–1949, A. Hermann, K. v. Meyenn and V. F. Weisskopf, eds. (Berlin: Springer, 1985), p. 516.

38. Letter from Pascual Jordan, Jr., to Daniela Wünsch, December 19, 1996, reported in Wünsch, "Theodor Kaluza," vol. 2, p. 111.

39. Schucking, "Jordan, Pauli, Politics, Brecht," p. 28.

40. W. H. McCrea, "Jordan's Cosmology," *Nature* 172, no. 4366 (July 4, 1953): 3–4.

41. Hans Reichenbach, *The Philosophy of Space and Time* (New York: Dover, 1957), pp. 281–282.

42. F. Rahaman, S. Chakraborty, N. Begum, M. Hossain, and M. Kalam, "A Study of Four and Higher-Dimensional Cosmological Models in Lyra Geometry," *Fizika B* 11 (2002): 57.

43. Dieter von Reeken, personal communication, December 16, 2002.

44. Letter from Wolfgang Pauli to Pascual Jordan, March 5, 1952, reprinted in Pauli, *Wissenschaftlicher Briefwechsel, (Scientific Correspondence),* vol. 4, part 1, 1950–1952, p. 568.

45. John R. Smythies, "Minds and Higher Dimensions," *Journal of the Society for Psychical Research* 55, no. 812 (1952).

46. John R. Smythies, personal communication, December 20, 2002.

47. John R. Smythies, "Space, Time and Consciousness," *Journal of Consciousness Studies* 10, no. 3 (2003): 47–56.

48. Lovecraft, "The Dunwich Horror," in *The Best of H.P. Lovecraft* (New York: Ballantine Books, 1982), p. 132.

49. H. P. Lovecraft, "Through the Gates of the Silver Key," *Weird Tales* 24, no. 1 (July 1934): 60.

50. H. P. Lovecraft, "The Dreams in the Witch House," in *The Best of H.P. Lovecraft*, p. 297.

51. Madeleine L'Engle, *A Wrinkle in Time* (New York: Scholastic Books, 1962), p. 94.

10   Gauging the Weak and the Strong

1. John Archibald Wheeler with Kenneth Ford, *Geons, Black Holes and Quantum Foam: A Life in Physics* (New York: Norton, 1998), p. 168.

2. Robert P. Crease and Charles C. Mann, *The Second Creation* (New York: Collier Books, 1986), p. 138.

3. Hans Bethe, "Quantum Theory," *Reviews of Modern Physics* 71, no. 2 (1999): S4.

4. Phone interview with Engelbert Schucking, December 9, 2002.

5. Joshua Goldberg, "Peter Bergmann" (talk presented at the American Physical Society, April 5, 2003).

6. Ernest Bergmann, personal communication, December 25, 2002.

7. Albert Einstein, report to the National Science Foundation, April 18, 1954.

8. Jagdish Mehra and Kimball A. Milton, *Climbing the Mountain: The Scientific Biography of Julian Schwinger* (Oxford: Oxford University Press, 2000), p. 640.

9. Phone interview with Bryce DeWitt, December 4, 2002.

10. Ibid.

11. Interview with Stanley Deser, Brandeis University, November 22, 2002.

12. Ibid.

13. Interview with John Wheeler, Princeton University, November 5, 2002.

14. Albrecht Fölsing, *Albert Einstein,* Ewald Osers, trans. (New York: Penguin, 1997), pp. 739–40.

15. Daniela Wünsch, "Theodor Kaluza: Leben und Werk (Life and Work)," vol. 2 (PhD diss., University of Stuttgart, 2000), pp. 140–41.

16. Letter from Helen Dukas to Abraham Pais, April 30, 1955, quoted in Abraham Pais, *Subtle Is the Lord: The Science and the Life of Albert Einstein* (New York: Oxford University Press, 1982), p. 332.

17. Interview with Stanley Deser, Brandeis University, November 22, 2002.

18. Stanley Deser, "Oskar Klein: From His Life and Physics," in *Proceedings of the Oskar Klein Centenary Symposium,* Ulf Lindström, ed. (Singapore: World Scientific, 1995), p. 50.

19. Interview with Stanley Deser, Brandeis University, November 22, 2002.

20. S. Goudsmit, editorial, *Physical Review Letters* 3, no. 11 (1959): 505.

21. Phone interview with Bryce DeWitt, December 4, 2002.

22. Ibid.

23. Lecture Notes by Bryce DeWitt, 1963, Les Houches Summer School, reprinted in Thomas Appelquist, Alan Chodos, and Peter Freund, *Modern Kaluza-Klein Theories* (Reading, Mass.: Addison-Wesley, 1987), p. 114.

24. Phone interview with Bryce DeWitt, December 4, 2002.
25. Interview with Stanley Deser, Brandeis University, November 22, 2002.
26. Deser, "Oskar Klein," p. 53.
27. Letter from Abdus Salam to Oskar Klein, January 18, 1965 (Niels Bohr Archive).
28. Letter from Oskar Klein to Abdus Salam, June 12, 1969 (Niels Bohr Archive).
29. Michael J. Duff, "A Tribute to Abdus Salam" (talk presented at the Workshop on Frontiers in Field Theory, Quantum Gravity and String Theory, Puri, India, December 12–21, 1996).

11 Hyperspace Packages Tied Up in Strings

1. Howard Georgi, quoted in Robert P. Crease and Charles C. Mann, *The Second Creation* (New York: Collier Books, 1986), p. 417.
2. Sheldon Glashow, interview on "NOVA: What Einstein Never Knew," originally broadcast October 22, 1985.
3. Claud Lovelace, personal communication, July 17, 2003.
4. John H. Schwarz, "Reminiscences of Collaborations with Joël Scherk" (talk presented at Conférence anniversaire du LPT-ENS (2000), Caltech preprint, p. 5.
5. Claud Lovelace, personal communication, July 17, 2003.
6. Interview with Bernard Julia, ENS, Paris, January 13, 2003.
7. Interview with Eugene Cremmer, ENS, Paris, January 13, 2003.
8. "Edward Witten," *Brandeis Review* (Summer 1999): 26.
9. Interview with Bernard Julia, ENS, Paris, January 13, 2003.
10. Bernard Julia, personal communication, July 30, 2003.
11. Interview with Bernard Julia, ENS, Paris, January 13, 2003.
12. Interview with Stanley Deser, Brandeis University, November 22, 2002.
13. Interview with Bernard Julia, ENS, Paris, January 13, 2003.
14. Interview with Stanley Deser, Brandeis University, November 22, 2002.
15. Interview with Bernard Julia, ENS, Paris, January 13, 2003.
16. Interview with Stanley Deser, Brandeis University, November 22, 2002.
17. John H. Schwarz, "Reminiscences of Collaborations with Joël Scherk," p. 5.
18. Interview with Eugene Cremmer, ENS, Paris, January 13, 2003.
19. Michio Kaku, *Hyperspace: A Scientific Odyssey through Parallel Universes, Time Warps and the Tenth Dimension* (New York: Oxford University Press, 1994), pp. 104–5.
20. Phone interview with Alan Chodos, September 11, 2002.
21. Edward Witten, "Search for a Realistic Kaluza-Klein Theory," reprinted in Thomas Appelquist, Alan Chodos, and Peter Freund, *Modern Kaluza-Klein Theories* (Reading, Mass.: Addison-Wesley, 1987), p. 282.
22. Barry Parker, *The Search for a Supertheory: From Atoms to Superstrings* (New York: Plenum, 1987), p. 247.

23. Michael J. Duff, "A Layman's Guide to M-Theory" (talk presented at the Abdus Salam Memorial Meeting, ICTP, Trieste, Italy, November 1997).

24. Sheldon Glashow, in *Seventh Workshop on Grand Unification,* J. Arafune, ed. (Philadelphia: World Scientific, 1987), p. 548.

25. Richard Feynman, in *Superstrings: A Theory of Everything?* P.C.W. Davies and J. Brown eds., (New York: Cambridge University Press, 1988), p. 193.

26. Albert Einstein, quoted by Peter G. Bergmann, "The Quest for Unity: General Relativity and Unitary Field Theories," in *To Fulfill a Vision: Jerusalem Einstein Centennial Symposium on Gauge Theories and Unification of Physical Forces,* Yuval Ne'eman, ed. (Reading, Mass.: Addison-Wesley, 1981), p. 20.

27. Michael J. Duff, "A Layman's Guide to M-Theory."

28. Ibid.

12   Brane Worlds and Parallel Universes

1. Kaey Davidson, "Parallel Universe Theory Not Just Sci-Fi," *San Francisco Examiner,* July 24, 2000, p. A1.

2. Savas Dimopoulos, personal communication, December 16, 2002.

3. Jeffrey Harvey, quoted in George Johnson, "Almost in Awe, Physicists Ponder 'Ultimate' Theory," *New York Times,* September 22, 1998, p. D1.

4. Ibid.

5. Burt Ovrut, quoted in George Johnson, "A Passion for Physical Realms, Minute and Massive," *New York Times,* February 20, 2001, p. D1.

6. Phone interview with Raman Sundrum, December 3, 2002.

7. Interview with Gary Gibbons, Cambridge University, January 15, 2003.

8. Ibid.

9. Interview with Lisa Randall, Harvard University, November 22, 2002.

10. Interview with Paul Steinhardt, Princeton University, November 5, 2002.

11. Interview with Lisa Randall, Harvard University, November 22, 2002.

12. Phone interview with Raman Sundrum, December 3, 2002.

13. Savas Dimopoulos, personal communication, December 16, 2002.

14. Maria Spiropulu, "In Search of Extra Dimensions" (talk presented at the AAAS Annual Meeting, Boston, February 17, 2002).

Conclusion: Extra-dimensional Perception

1. Joseph Lykken, "Mysteries of Extra Dimensions" (talk presented at the American Physical Society Annual Meeting, Philadelphia, April 5, 2003).

2. Phone interview with Alan Chodos, September 11, 2002.

3. Ibid.
4. Tony Robbin, *Fourfield: Computers, Art and the Fourth Dimension* (Boston: Bulfinch Press, 1992, p. 25).
5. Peter Rose, personal communication, December 15, 2002.
6. Rudy Rucker, foreword, in Robbin, *Fourfield,* p. 11.
7. Michael J. Duff, "The World in Eleven Dimensions: A Tribute to Oskar Klein," Oskar Klein Professorship Inaugural Lecture, University of Michigan, March 16, 2001.

Further Reading

(Technical works are marked with an asterisk.)

Abbott, Edwin Abbott. *Flatland: A Romance of Many Dimensions.* London: Seeley and Co., 1884.

———. *The Annotated Flatland.* Introduction and notes by Ian Stewart. Cambridge, Mass.: Perseus, 2002.

*Appelquist, Thomas, Alan Chodos, and Peter G. O. Freund. *Modern Kaluza-Klein Theories.* New York: Addison-Wesley, 1987. This collection includes many of the most important papers in Kaluza-Klein theory, including English translations of the following seminal works:

*Kaluza, Theodor. "Zur Unitätsproblem der Physik." *Sitzungsberichte der Preussischen Akademie der Wissenschaften* 54 (1921): 966–72.

*Klein, Oskar. "Quantentheorie und fünfdimensionale Relativitätstheorie." *Zeitschrift für Physik* 37 (1926): 895.

Arkani-Hamed Nima, Dimopoulos Savas, and Dvali Georgi. "The Universe's Unseen Dimensions." *Scientific American* 283, no. 2 (August 2000): 62–69.

Banchoff, Thomas F. *Beyond the Third Dimension: Geometry, Computer Graphics and Higher Dimensions.* New York: Scientific American Library, 1990.

Barrow, John. "Dimensionality." *Philosophical Transactions of the Royal Society of Astronomy* 310 (1983): 337–46.

Bell, Eric Temple. *Men of Mathematics.* New York: Simon and Schuster, 1937.

Bergmann, Peter. *Introduction to the Theory of Relativity.* New York: Prentice-Hall, 1942.

———. *The Riddle of Gravitation.* New York: Charles Scribner's Sons, 1968.

Bonola, Roberto. *Non-Euclidean Geometry: A Critical and Historical Study of Its Development.* Chicago: Open Court, 1912.

Boyer, Carl B. *A History of Mathematics.* New York: John Wiley & Sons, 1968.

Bragdon, Claude Fayette. *A Primer of Higher Space.* Rochester, N.Y.: Manas Press, 1913.

———. *Four-Dimensional Vistas.* New York: Alfred A. Knopf, 1916.

*Cayley, Arthur. "Chapters in the Analytical Geometry of n Dimensions." *Cambridge Mathematical Journal* (1843).

*Chodos, Alan, and Steven Detweiler. "Where Has the Fifth Dimension Gone?" *Physical Review* D21 (1980): 2167–70.

Clark, Ronald W., *Einstein: The Life and Times*. New York: Avon Books, 1971.

Clifford, William Kingdon. "On the Space-Theory of Matter." *Proceedings on the Cambridge Philosophical Society* 2 (1876): 157–58.

Crease, Robert P., and Charles C. Mann. *The Second Creation*. New York: Collier Books, 1986.

Davies, Paul. *Superforce: The Search for a Grand Unified Theory of Nature*. New York: Simon and Schuster, 1984.

Davies, Paul, and Julian Brown, eds. *Superstrings: A Theory of Everything?* Cambridge: Cambridge University Press, 1988.

*De Sabbata, Venzo, and Ernst Schmutzer, eds. *Unified Field Theories of More than Four Dimensions: Proceedings of the International School of Cosmology and Gravitation*. Singapore: World Scientific, 1983. This contains English translations of some of the correspondence between Kaluza and Einstein.

Deutsch, A. J. "A Subway Named Mobius." *Astounding Science Fiction*. December 1950.

Duff, Michael J. "The Theory Formerly Known as Strings." *Scientific American* 2 (February 1998): 64–69.

*Duff, Michael J., ed. *The World in Eleven Dimensions: Supergravity, Supermembranes and M-theory*. Philadelphia: Institute of Physics, 1999.

Eddington, Arthur. *The Mathematical Theory of Relativity*. New York: Chelsea, 1924.

Ehrenfest, Paul. "In What Way Does It Become Manifest in the Fundamental Laws of Physics that Space Has Three Dimensions?" *Proceedings of the Royal Academy* (Amsterdam) 20 (1917): 200–209.

Einstein, Albert. *Autobiographical Notes*. Paul Arthur Schilpp, ed. and trans. La Salle, Ill.: Open Court, 1979.

———. *Ideas and Opinions*. Sonja Bargmann, trans. New York: Bonanza Books, 1954.

———. *The Meaning of Relativity*. Princeton, N.J.: Princeton University Press, 1956.

*Einstein, Albert, and Peter Bergmann. "On a Generalization of Kaluza's Theory of Electricity." *Annals of Mathematics* 39 (1938): 683–701.

Fölsing, Albrecht. *Albert Einstein*. Ewald Osers, trans. New York: Penguin, 1997.

Galison, Peter. "Minkowski's Spacetime: From Visual Thinking to the Absolute World." *Historical Studies in the Physical Sciences*, 10 (1979): 85–121.

Gamow, George. *One Two Three . . . Infinity*. New York: Viking, 1947.

Gardner, E. L. "The Fourth Dimension." *Theosophist* (October 1916): 53–63.

Gibbons, Gary. "Brane-Worlds." *Science* 287 (January 7, 2000): 49–50.

Goenner, Hubert. "Unified Field Theories: From Eddington and Einstein up to Now." In *Proceedings of the Sir Arthur Eddington Centenary Symposium,* vol. 1. Edited by V. de Sabbata and T. M. Karade. Singapore: World Scientific, 1984, pp. 176–96.

Golding, John. *Marcel Duchamp: The Bride Stripped Bare by Her Bachelors, Even.* New York: Viking Press, 1973.

Gray, Jeremy. *Ideas of Space.* Oxford: Oxford University Press, 1989.

Green, Michael. "Superstrings." *Scientific American,* September 1986.

Greene, Brian. *The Elegant Universe: Superstrings, Hidden Dimensions, and the Quest for the Ultimate Theory.* New York: Vintage Books, 2000.

Halpern, Paul. *Cosmic Wormholes: The Search for Interstellar Shortcuts.* New York: Dutton, 1992.

———. *The Pursuit of Destiny: A History of Prediction.* Cambridge, Mass.: Perseus, 2000.

———. *Time Journeys: A Search for Cosmic Destiny and Meaning.* New York: McGraw-Hill, 1990.

Heinlein, Robert. "And He Built a Crooked House." *Astounding Science Fiction,* February 1941.

Henderson, Linda. *The Fourth Dimension and Non-Euclidean Geometry in Modern Art.* Princeton, N.J.: Princeton University Press, 1983.

Hinton, Charles. *An Episode of Flatland, or How a Plane Folk Discovered the Fourth Dimension.* London: Sonnenschein, 1907.

———. *The Fourth Dimension.* New York: Arno Press, 1976.

———. "What Is the Fourth Dimension?" Reprinted in *Speculations of the Fourth Dimension: Selected Writings of Charles H. Hinton,* edited by Rudolf v. B. Rucker. New York: Dover, 1980, pp. 1–22.

Hoffmann, Banesh. *Relativity and Its Roots.* New York: Scientific American Books, 1983.

Hoffmann, Banesh, with Helen Dukas. *Albert Einstein: Creator and Rebel.* New York: Viking, 1972.

Holton, Gerald, and Yehuda Elkana, eds. *Albert Einstein: Historical and Cultural Perspectives.* Princeton, N.J.: Princeton University Press, 1982.

*Horava, Petr, and Edward Witten. "Eleven-Dimensional Supergravity on a Manifold with Boundary." *Nuclear Physics B475* (1996): 94–114.

Jammer, Max. *Concepts of Space: The History of Theories of Space in Physics.* Cambridge, Mass.: Harvard University Press, 1954.

Kaku, Michio. *Hyperspace: A Scientific Odyssey Through Parallel Universes, Time Warps, and the 10th Dimension.* New York: Doubleday, 1994.

Kant, Immanuel. *Critique of Pure Reason.* Translated by Norman Kemp Smith. London: Macmillan, 1929.

Klein, Martin J. *Paul Ehrenfest: The Making of a Theoretical Physicist.* London: North-Holland, 1970.

Klein, Oskar. "From a Life of Physics." In *From My Life of Physics.* Edited by Abdus Salam et al. Singapore: World Scientific, 1989.

Kline, Morris. *Mathematics in Western Culture.* New York: Oxford University Press, 1953.

Lanczos, Cornelius. *Space through the Ages: The Evolution of Geometrical Ideas from Pythagoras to Hilbert and Einstein.* New York: Academic Press, 1970.

Laporte, Paul M. "The Space-Time Concept in the Work of Picasso." *Magazine of Art* 41 (January 1948): 26–32.

*Lee, H. C., ed. *An Introduction to Kaluza-Klein Theories.* Singapore: World Scientific, 1984.

L'Engle, Madeleine. *A Wrinkle in Time.* New York: Yearling Books, 1973.

Lovecraft, H. P. "The Dreams in the Witch House." In *The Best of H. P. Lovecraft.* New York: Ballantine Books, 1982.

Manning, Henry. *The Fourth Dimension Simply Explained.* New York: Peter Smith, 1941.

———. *Geometry of Four Dimensions.* New York: Macmillan, 1914.

Mlodinow, Leonard. *Euclid's Window: The Story of Geometry from Parallel Lines to Hyperspace.* New York: Simon and Schuster, 2001.

Neville, Eric. *The Fourth Dimension.* Cambridge: Cambridge University Press, 1921.

Newcomb, Simon. "The Fairyland of Geometry." *Harper's Monthly* 104 (January 1902): 249–52.

———. "Modern Mathematical Thought." *Nature* 49 (February 1, 1894): 325–29.

———. "The Philosophy of Hyperspace." *Bulletin of the American Mathematical Society* 4 (February 1898): 187–95.

Pais, Abraham. *Subtle Is the Lord . . . : The Science and the Life of Albert Einstein.* Oxford: Oxford University Press, 1982.

Parker, Barry. *Einstein's Dream: The Search for a Unified Theory of the Universe.* New York, Plenum, 1986.

———. *Search for a Supertheory: From Atoms to Superstrings.* New York: Plenum, 1987.

Pickover, Clifford. *Surfing through Hyperspace: Understanding Higher Universes in Six Easy Lessons.* New York: Oxford University Press, 1999.

Poincaré, Henri. *The Value of Science.* New York: Modern Library, 2001.

Putz, John F. "Going Out on a Limb: A Reading and Writing Course about the Fourth Dimension." *Primus* 11 (2001): 1–15.

*Randall, Lisa, and Raman Sundrum. "An Alternative to Compactification." *Physical Review Letters* 83 (1999): 4690–93.

Rashevsky, Nicholas P. "Is Time the Fourth Dimension?" *Scientific American* 131 (December 1924): 400–402, 446.

Reichenbach, Hans. *The Philosophy of Space and Time.* Translated by Maria Reichenbach and John Freund. New York: Dover, 1957.

Reid, Constance. *Hilbert.* Berlin: Springer, 1970.

Riemann, Bernard. "On the Hypotheses Which Lie at the Bases of Geometry." William Kingdon Clifford, trans. *Nature* 8 (May 1, 1873): 14–17.

Richardson, John Adkins. *Modern Art and Scientific Thought.* Urbana: University of Illinois Press, 1971.

Robbin, Tony. *Fourfield: Computers, Arts and the Fourth Dimension.* Boston: Bulfinch Press, 1992.

———. "The New Art of Four-Dimensional Space: Spatial Complexity in Recent New York Work." *Artscribe,* no. 9 (1977): 19–22.

Rucker, Rudy. *The Fourth Dimension.* Boston: Houghton Mifflin, 1984.

———. *Geometry, Relativity and the Fourth Dimension.* New York: Dover, 1977.

Shlain, Leonard. *Art and Physics: Parallel Visions in Space, Time and Light.* New York: William Morrow, 1991.

Smolin, Lee. *Three Roads to Quantum Gravity.* New York: Basic Books, 2001.

Stachel, John. *Einstein from "B" to "Z."* Boston: Birkhäuser, 2002.

———. "History of Relativity." In *Twentieth Century Physics,* vol. 1. Laurie Brown et al., eds. New York: American Institute of Physics Press, 1995.

Stringham, W. I. "Regular Figures in n-Dimensional Space." *American Journal of Mathematics* 3 (1880): 1–12.

Sylvester, James Joseph. "Four-Dimensional Space," *Nature* 31 (March 1885): 481.

Tegmark, Max. "On the Dimensionality of Spacetime." *Classical and Quantum Gravity* 14 (1997): L69–L75.

Van Nieuwenhuizen, Peter, and Daniel Z. Freedman. "The Hidden Dimensions of Spacetime." *Scientific American,* March 1985.

Van Oss, Rocine. "D'Alembert and the Fourth Dimension." *Historia Mathematica* 10 (November 1983): 455–57.

*Vizgin, Vladimir. "The Geometrical Unified Field Theory Program." In *Einstein and the History of General Relativity.* Edited by Don Howard and John Stachel. Boston: Birkhäuser, 1989, pp. 300–314.

Weinberg, Steven. *Dreams of a Final Theory: The Scientist's Search for the Ultimate Laws of Nature.* New York: Vintage, 1992.

Wells, H. G. "The Plattner Story." *New Review* (April 1896).

———. *The Time Machine, an Invention.* London: W. Heinemann, 1895.

Wesson, Paul S. *Space-Time-Matter: Modern Kaluza-Klein Theory.* Singapore: World Scientific, 1999.

Weyl, Hermann. *Space, Time, Matter.* New York: Dover, 1950.

Wheeler, John Archibald, with Kenneth Ford. *Geons, Black Holes and Qunatum Foam: A Life in Physics.* New York: Norton, 1998.

Whitrow, G. J. "Why Physical Space has Three Dimensions." *British Journal for the Philosophy of Science* 6 (1955): 13–31.

*Witten, Edward. "Search for a Realistic Kaluza-Klein Theory." *Nuclear Physics* B186 (1981): 412–28.

Wünsch, Daniela. "Theodor Kaluza: Leben und Werk (Life and Work)." Ph.D. diss. University of Stuttgart, 2000.

Zöllner, Johann Carl Friedrich. "On Space of Four Dimensions." *Quarterly Journal of Science* 8 (April 1878): 227–37.

Index

Abbott, Edwin Abbott, 50, 53–57, 82, 93, 101, 275, 295
ADM formalism, 220
Age of Exploration, 17
Albrecht, Andreas, 283
Alfvén, Hannes, 228
Allegory of the Cave (Plato), 12–13, 187
Alvarez-Gaumé, Luis, 249
Ampère, André-Marie, 62
Anderson, Jim, 219
antigravity, 249, 272
anti-matter annihilation, 228
antineutrinos, 248
antiparticles, 223
antiquarks, 224, 232
Antoniadas, Ignatius, 269, 270
Appelquist, Thomas, 246, 291
Aristotle, 13
Arkani-Hamed, Nima, 268–271, 272, 274, 276, 286
Arnowitt, Richard, 214, 220
Arrhenius, Svante, 115, 116, 118, 120
art
    four-dimensional, 81–82, 83
    higher dimensions, 293–294
    three-dimensional, 15–16
    two-dimensional, 15
Ashtekar, Abhay, 258, 259
astral plane, fourth dimension, 46
astronomy, Galileo Galilei, 17
atomic bomb. *See also* nuclear physics
    Bergmann and, 190
    Germany and, 196
    Nazi Germany and, 178

background radiation, 284
Bacon, Francis, 54
Balla, Giacomo, 81
Banchoff, Thomas, 47, 293–294, 295
Bargmann, Sonja, 191
Bargmann, Valentine, 163, 166–171, 174, 180, 191, 193, 196, 217, 238, 247, 291
Bartels, Johann, 29
baryonic matter, 272
baryons, 223, 224
baseball gun, 52
Becquerel, Henri, 2, 176
Beichler, James, 57
Bergmann, Emmy, 164, 165
Bergmann, Ernest, 189, 212
Bergmann, Esther, 164
Bergmann, Margot (née Eisenhardt), 164, 189, 190, 257
Bergmann, Martha, 164
Bergmann, Max, 163, 164, 167, 184
Bergmann, Peter, 163–171, 169, 170, 171, 174, 178, 180, 188–191, 193, 198–199, 206,

210–212, 213, 214, 217, 219, 220, 221, 240, 257–258, 259, 274, 291
Bergshoeff, Eric, 261
Berryman, Clifford, 8
Besso, Michele, 69, 87, 144
beta decay, 2
Bethe, Hans, 210
Beyer, Anna (Anna Kaluza), 104, 112
Big Bang theory, 228, 246, 272, 283, 284–285
Big Crunch, 284
black holes, 277, 282, 284
Blavatsky, Helena P., 45–46
Bohr, Niels, 95, 113, 115, 119, 120–121, 122, 124, 127, 130, 131, 132, 133, 135, 136, 137, 142, 144, 145–146, 148, 151, 156, 157, 162, 169, 171, 175, 177, 180, 183, 196, 207, 209, 278
Bolyai, János, 30–31, 35, 36, 37
Bolyai, Wolfgang, 14, 30
Boole, George, 50
Born, Max, 116, 132, 136, 145, 146, 184, 217
bosons, 222, 232, 234, 235, 236, 265
Bragdon, Claude, 26, 45, 46
brane models, 9, 11, 12, 195, 202, 217, 290. *See also* string theory; superstring theory
    collision scenario, 284
    cosmology, 285
    dark matter, 272–273
    gravity, 269–271
    higher dimensions, 282
    Kaluza-Klein theory, 276
    Maldacena, 277–278
    parallel universe, 271–272
    Randall-Sundrum, 280–281
    string theory, 264–266
Brans-Dicke scalar (dilation), 109
Braque, Georges, 81
Broad, C. D., 202
Broglie, Louis de, 123–124, 126, 129, 131, 144, 146, 252
Bruno, Giordano, 17
bulk, 11, 12
Bunsen, Robert Wilhelm, 116
Burger, Dionys, 57

Calabi, Eugenio, 256
Calabi-Yau spaces, 256, 264, 293
Candelas, Philip, 255
Carr, Bernard, 202
Casimir effect, 246, 280
cave allegory (Plato), 12–13, 187
Cayley, Arthur, 39, 40, 41, 42, 49, 55, 65, 292
CERN accelerator (Switzerland), 227, 232, 233, 239, 241, 286
Chadwick, James, 176
Chaplin, Charlie, 139–144, 147, 148, 150, 155, 200
Chaucer, Geoffrey, 103
chirality (handedness), 21–24, 248–250
Cho, Yong Min, 245
Chodos, Alan, 245–246, 291
Christianity, Middle Ages, 15
Clifford, William Klingdon, 38–40, 42, 43, 47, 55, 65, 91, 111, 219

Cohen, Andrew, 274
Colby, Walter, 123
cold dark matter, 272
Coleman, Sidney, 238
collective unconscious, 192, 202
Como Lecture, 145–146
compactification, 282
complementarity principle, 120, 133, 135, 145
Compton effect, 137
computer graphics, hyperspace representations, 47
connections, general relativity, 88
Copernican universe, 17
cosmic background radiation, 284
cosmological constant, 249, 280
cosmology, 217, 228, 246, 272, 275, 279, 280, 283–289. See also universe
Coulomb, Charles-Augustin, 62
Courant, Richard, 28, 184, 185
Cremmer, Eugene, 230, 242, 243, 244, 248, 249
Crookes, William, 46, 201
cube, 13
Cubism, 81, 83
curvature concept, Riemannian geometry, 36–37
curvature tensor, 88, 109
Cyclic Universe, 284, 285
cylinder condition, 109–110, 113

D'Alembert, Jean, 18, 19, 24, 57, 71
Dali, Salvador, 51, 52
dark energy, 272
dark matter, brane models, 272–273
Darwin, Charles, 117, 292
d'Auria, Riccardo, 250
Davies, Marion, 155
da Vinci, Leonardo, 16, 81, 83
Debye, Peter, 119
Deconstructing Dimensions project, 274–276
Derrida, Jacques, 274
Deser, Elsbeth (née Klein), 183, 214
Deser, Stanley, 6, 181–182, 191–192, 213–214, 218, 219–220, 221, 227, 228, 239, 241, 243, 249
de Sitter, Willem, 152, 204, 280, 281, 282, 283
Detweiler, Steven, 245–246
Deutsch, A. J., 204, 295
Dewdney, A. K., 57
de Wit, Bernard, 249
DeWitt, Bryce Seligman, 212–213, 214, 219, 220, 221, 240, 245
DeWitt-Morette, Cécile, 214, 219, 220–221
Dickens, Charles, 103
Diderot, Denis, 18
dilaton (Brans-Dicke scalar), 109
dimensionality. See also higher dimensions; specific dimensions
   of objects and space, 261
   researches in, 38–40
   supersymmetry, 233
Dimopoulos, Savas, 269, 270–271, 272, 286, 287
Dirac, Paul, 115, 116, 135, 136, 137, 180, 197, 198, 219, 220, 223, 233, 238, 240, 257, 261, 269

Dirac equation, 137, 208, 238
distant parallelism, 149, 150, 193
dodecahedron, 13
duality relationships, 261–262, 263
Duchamp, Marcel, 81–82, 83
Duff, Michael J., 230, 249, 261, 262, 296
Dukas, Helen, 150, 156, 169
Dunne, J. W., 141
Dvali, Georgi (Gia), 269, 270–271, 272, 278, 286
Dyson, Freeman, 193, 210

Eastman, George, 160
Eddington, Arthur, 111, 112, 135, 148, 149, 152, 180, 193
Ehrenfest, Paul, 86, 92–96, 100, 102, 105, 113, 120, 121–122, 129–130, 131, 134–135, 144, 145, 146, 156–157, 193, 271
Ehrenfest, Tatyana, 93–94, 105, 156
Ehrenfest, Vassik, 157
Einstein, Albert, 1, 2, 5, 6, 7–9, 18, 25, 33, 35, 57, 60, 61, 68–71, 72, 79, 80–81, 82, 83, 84–92, 94, 96–100, 107, 108, 109, 110–111, 112–113, 119, 121, 122, 125, 126, 129, 131, 132, 135, 138, 139, 140–163, 165–175, 180, 184, 185, 188–189, 191, 192, 193–195, 198, 204, 206, 207, 210, 211–212, 213, 215–217, 226, 233, 238, 246, 249, 256, 258, 260, 267, 274, 280, 282, 288, 291, 296
Einstein, Elsa (née Löwenthal), 86–87, 96, 140, 141, 149, 153, 156, 169
Ekpyrotic Universe, 284, 285
electromagnetism
   Einstein, 172, 174–175
   interactions of, 61–62
   Kaluza, 108–109
   Kaluza-Klein theory, 5
   Klein, 125–126
   natural forces, 2, 3, 4
   nuclear physics, 176
   research in, 63–66, 72
   Weyl, Hermann, 96–99
electroweak theory, 4, 234, 240, 247, 254
eleven-dimensional supergravity, 218, 242–244, 245, 248, 251, 261
embedding concept, Riemannian geometry, 36–37
Englert, François, 249
Eöt-Wash group, 287
EPR paradox, 162
equivalence principle, 84–85
Euclid, 13–14, 16, 24, 25, 31, 36
Euler, Leonhard, 103
Euler's theorem, 47
evolution theory, 58
extrasensory perception (ESP), 201

Fairbanks, Douglas, 139
Faraday, Michael, 62, 63–64
fascism, 135, 155
Ferdinand, Carl Wilhelm, 29–30
Fermi, Enrico, 176
fermions, 222, 232, 234, 236, 248
Ferrara, Sergio, 239
Feynman, Richard, 206, 207–210, 219, 220, 239, 250, 252, 253, 256–257, 260

fifth dimension. *See also* higher dimensions;
    Kaluza-Klein theory
    Bergmann, 189
    Einstein, 148, 151, 171, 173–174, 195
    Jordan, 199
    Julia, 238
    Kaluza, 107–113, 187
    Kaluza-Klein theory, 6–7, 8
    Klein, 124–127, 128–129, 130–131, 136,
        137, 174, 178, 180–183, 229
    Nordström, 91–92
    Pauli, 193, 195
    projective relativity, 159
    Thiry, 199
Fischer, Emil, 163
Fitzgerald, George, 68, 71, 78
Flexner, Abraham, 152, 153, 155, 159, 161,
    167
Flexner, Simon, 167
Fock, Vladimir, 129
Fokker, Adriaan, 87
Fomalont, Ed, 288
Fontenelle, Bernard, 17
Ford, Ken, 191
fourth dimension, 26
    art, 81–82
    Bohr, 121
    concepts of, 19–21
    Einstein, 80–81
    essays on, 82–83
    handedness (chirality), 21–24
    identification of, 18–19
    Minkowski, 73–74
    Newtonian physics, 18
    religion, 44
    space-time, 73–78
    time, 18–19, 57–60, 71, 73
Fraenkel, Adolf, 113
Franck, James, 184
Frank, Philipp, 165, 167, 188
Fré, Pietro, 250
Frederick the Great (king of Prussia), 103
Freedman, Daniel, 239
Freud, Sigmund, 192
Freund, Peter, 91, 244–245, 246
Futurism, 81–82

Galileo Galilei, 17, 88, 228
Galison, Peter, 72
Gamow, George, 15, 137, 180, 294
Gates, S. James, Jr., 250
gauge, 97
gauge theory, 182, 201, 221, 225, 232, 234,
    245, 254, 276, 292
Gauss, Carl Friedrich, 14, 24–25, 28–30, 31,
    32, 33, 34–35, 38, 39, 42, 43, 72, 82, 104,
    203
Gell-Mann, Murray, 222, 223, 224, 228, 251
general theory of relativity, 40, 42
    Einstein, 84–90, 172
    gravity, 2
    Kaluza, 107–108, 109
    Kaluza-Klein theory, 5, 8
    Kasner solution, 246
    quantum theory and, 211–212, 258
    research in, 212–214, 219–220

Riemann, 25, 35
scalar-tensor theory (Jordan-Brans-Dicke
    model), 198–199
geometry
    Euclidean, 13–14, 29, 36
    non-Euclidean, 24–25, 29, 30–32, 35–37, 40,
        86
    quantum theory, 258–259
    unified theory, 282
geon theory, 219
Georgi, Howard, 227, 231, 269, 274, 276, 279
Gibbons, Gary, 249, 282–283
Giddings, Steven, 278
Giotto di Bondone, 15, 81
Glashow, Sheldon, 4, 224, 226, 227, 230,
    231–232, 248, 263, 268
gluons, 265
Goethe, Johann Wolfgang von, 27
Goldberg, Joshua, 211, 214
Gordon, E. A. Hamilton, 58
Gordon, Walter, 115, 136, 183, 184
Göttingen, Germany, 27–28
Goudsmit, Samuel, 130, 219
Grand Unified Theory (GUT), 227, 269
Grassmann, Hermann, 39
gravitons, 265, 270, 289
gravity
    brane models, 269–271
    dimensionality of space and, 20
    Einstein, 84–90, 172, 174–175, 216, 235
    higher dimensions, 282
    interactions of, 62
    Kaluza-Klein theory, 4–9
    M-theory, 12
    natural forces, 2, 3, 4
    Newton, 17–18
    non-Euclidean geometry, 38
    quantum electrodynamics (QED), 210–212
    quantum theory, 240, 258–259
    researches in, 214, 217–220, 235–239
    scalar-tensor theory (Jordan-Brans-Dicke
        model), 198–199
    special theory of relativity, 85
    supersymmetry, 235
    testing of, 287–289
    Weyl, 96–99
Great Depression, 139
Greece (ancient), 12–14
Green, Michael, 243–244, 250–251, 253–254
Greene, Brian, 296
Grimm, Jakob, 27–28
Grimm, Wilhelm, 27–28
Grommer, Jakob, 111, 113, 150
Gross, David, 255
Grossmann, Marcel, 69, 85, 86
ground state, 95
Grunwald, Clara, 164, 167
Gubser, Steven, 278
Guth, Alan, 283

hadronic string theory, 232–234
hadrons, 222–223
Hahn, Otto, 175, 183
hallucinogens, 202
Hamilton, William, 39, 63
Hamilton-Jacobi equation, 125

handedness (chirality), 21–24, 248–250
Harvey, Jeffrey, 255, 276, 277–278
Hasse, Helmut, 185, 186, 188
Hawking, Stephen, 285
Hearst, William Randolph, 139, 155
Heinlein, Robert, 204–205, 295
Heisenberg, Werner, 115, 116, 128, 130, 132,
     133, 136, 137, 142, 146–147, 176, 177,
     180, 184, 196, 204, 208, 211, 288
Helmholtz, Hermann Ludwig Ferdinand von,
     72, 82, 116, 118
Henderson, Linda Dalrymple, 82, 83
Hendrix, Jimi, 252
Henry, Joseph, 62
Hertz, Heinrich, 66, 72
Herzfeld, Karl, 113, 169
heterotic string theory, 255
hierarchy problem, 4, 269
Higgs, Peter, 225–226
Higgs mechanism, 240, 243
higher dimensions. See also fifth dimension;
     Kaluza-Klein theory
   Bohr, 121
   brane models, 282
   cosmology, 286–289
   Deconstructing Dimensions project,
     274–276
   Ehrenfest, 93–95
   Einstein, 141–142, 150–152, 171
   eleven-dimensional supergravity, 218,
     242–244, 245, 248
   gravity, 282
   hadronic string theory, 232–234
   Julia, 238
   Kaluza-Klein theory, 4–9
   mysticism, 44–46
   researches in, 38–40, 41, 43, 220–221
   Smythies, 202
   supersymmetry, 231–235
   teaching of, 292–296
   technology, 296–297
   testing of, 286–293
Hilbert, David, 72, 84, 93, 96, 104, 105, 136,
     184, 185
Hinton, Charles Howard, 49, 50–53, 55, 57,
     82, 93, 295
Hinton, James, 50
Hinton, Mary, 50, 51
Hitler, Adolf, 155, 157, 165, 178, 180, 183,
     188, 196. See also Nazi Germany; World
     War II
Hoffmann, Banesh, 158, 159, 160, 162, 193
holographic principle, 277–278
Holton, Gerald, 83, 179, 188
Horava, Petr, 264, 270
Horowitz, Gary, 255
hot matter, 284
Howe, Paul, 261
Hubble, Edwin, 138
Hughes, James, 261
Hugo, Victor, 241
Humboldt, Alexander von, 30
Huxley, Aldous, 202
Huxley, Thomas, 58
hydrodynamics, electromagnetism, 64
hypercube, 47

hyperspace. See also higher dimensions
   concept's origin, 34
   hierarchy problem, 4–9
   literature, 49–50, 54–57, 58–59
   perceptions of, 50–53
   representations of, 47–49, 187
   researches in, 38–43
   Riemannian geometry, 35–37
   supernatural and, 44–46
   teaching of, 292–296

Iacocca, Lee, 190
icosahedron, 13
Inami, Takeo, 261
Indonesian puppet theater (Wayang Kulit),
     10–11
Infeld, Leopold, 18, 151, 162, 214, 217
inflationary universe model, 283–285
Institute for Advanced Studies (Princeton),
     152–153, 155–156, 159, 166, 184,
     188–189, 193–196, 235
International Centre for Theoretical Physics
     (ICTP), 228–229
International Union of Physics, 180
isospin, 177, 181
Israeli-Palestinian conflict, 238

James, William, 49
Joos, Georg, 186
Jordan, Pascual, 115, 135, 136, 191, 196–199,
     200, 201, 206, 215, 217, 246, 269
Jordan-Brans-Dicke model (scalar-tensor the-
     ory), 198–199, 206, 215
Joseph, David W., 281–282
Jouffret, E., 83
Julia, Bernard, 230, 236, 238, 239, 240, 242,
     244, 248, 249
Jung, Carl, 192, 200, 202

Kaku, Michio, 296
Kaloper, Nemanja, 272
Kaluza, Amalie, 102
Kaluza, Anna, 104, 112
Kaluza, Augustin, 103
Kaluza, Dorothea, 102, 107, 186
Kaluza, Max, 102, 103
Kaluza, Theodor Franz Eduard, 1, 5–9, 25, 33,
     91, 99–100, 101–113, 115, 128, 129, 148,
     183–188, 189, 195, 196, 198–199,
     200–201, 215–216, 246, 291, 298
Kaluza, Theodor Franz Eduard, Jr., 107, 112
Kaluza-Klein theory, 40, 48, 51, 91, 230,
     255–256, 276, 282, 288, 289, 290, 291. See
     also fifth dimension; higher dimensions
   Bergmann, 189, 198
   cosmology, 246
   described, 4–9
   development of, 217–218, 220–221
   Einstein, 150, 151, 159, 160, 171, 174, 195,
     206, 216
   Freund, 244–245
   Hoffmann, 160–161
   Julia, 238
   legacy of, 297–298
   M-theory, 12
   Pauli, 193, 195

Smythies, 202
string theory, 260
subatomic physics, 206
supergravity, 243
Witten, 247–250
Kant, Immanuel, 20–21, 24, 43, 50, 72, 82, 95, 103
Kasner solution, 246
Kaufman, Bruria, 216, 217
Kent, Clark, 227
Khoury, Justin, 284
Kirchhoff, Gustav Robert, 116
Klein, Felix, 48, 49, 66, 91, 93, 94, 104, 105, 187
Klein, Gerda Agnete (née Koch), 123, 179
Klein, Gottlieb, 116, 118
Klein, Martin J., 94, 157
Klein, Oskar Benjamin, 5–9, 48, 113, 114–138, 148, 150, 171, 174, 178, 179–183, 191, 193, 214, 217, 218, 219, 228–230, 238, 255, 278, 291, 298
Klein bottle, 23, 48
Klein-Gordon equation, 136
Kneser, Martin, 106
Kopeikin, Sergei, 288
Kramers, Hendrik, 119, 120, 121, 128, 129, 180
Krauss, Lawrence, 280

Laemmle, Carl, 140
Lagrange, Joseph-Louis, 18–19, 57, 71, 241–242
Landsberg, Greg, 289
Large Hadron Collider (LHC), 286, 289, 292, 297
Laub, Jakob, 79
Laue, Max von, 80, 155, 216, 217
Leadbeater, C. W., 46
League of Nations, 180
Lee, T. D., 228, 248
Lenard, Philipp, 196
L'Engle, Madeleine, 205
leptons, 223
Levinson, Barry, 236–237
Levy, Antonie, 116
Lichnerowicz, André, 199, 218
light. See also optics
    Bohr, 119
    Einstein, 144
    shadow and, three-dimensional art, 16
    special theory of relativity, 69
    speed of, 66–68
Linde, Andrei, 202, 283, 285
Listing, Johann, 23, 33
literature
    fourth dimension, 82–83
    higher dimensions, 203–205
    hyperspace, 49–50, 54–57, 58–59
    parallel universes, 294–295
Liu, Jun, 261
Lobachevsky, Nikolai, 14, 31, 35, 36, 37
Long, Joshua, 288
loop quantum gravity, 258–260, 263
Lorentz, Hendrik, 68, 71, 72, 77, 78, 85, 92, 94, 118, 130, 146, 156
Lovecraft, H. P., 203–204

Lovelace, Claud, 233–234
Lykken, Joseph, 288, 290
Lyra, Gerhard, 200–201, 215

magnetism. See electromagnetism
Maldacena, Juan, 276–278, 280, 291
Mandel, Heinrich, 129, 137
Manhattan Project, Bergmann, 190
manifold concept, Riemannian geometry, 35
Manyfold Universe, 272–273, 297
Maric´, Mileva, 69, 86, 87, 96
Martinec, Emil, 255
Massive Compact Halo Objects (MACHOs), 272, 273
mathematics, three-dimensional geometry, 13–14
matrices, 39, 41, 181
Maxwell, James Clerk, 2, 5–6, 10, 62–67, 69, 71, 78, 87, 93, 97, 98, 108, 109, 125, 276
Mayer, F. W. F., 104
Mayer, Walther, 9, 149–150, 151, 153, 154, 156, 159, 161
McGovern, George, 238
media
    Einstein, 149
    higher dimensions, 202–205, 294–296
Meitner, Lise, 183
mescaline, 202
mesons, 177, 181, 183, 223, 224
metric concept, Riemannian geometry, 35–36
metric tensor, 77, 97, 108, 216
Michelangelo, 16, 83
Michelson, Albert, 67, 68
Middle Ages, 15, 16–17
Mie, Gustav, 87
Millikan, Robert, 138, 155
Mills, Robert, 182, 221, 223, 234
Milton, John, 54
Minkowski, Hermann, 57, 60, 71–80, 83, 84, 87, 88, 89, 90, 91, 104, 105
Misner, Charles, 214, 220
Möbius, August, 23, 24, 25
Möbius strip, 23, 24, 33, 204
Moller, Christian, 183
Montessori, Maria, 164
Montonen, Claus, 263
Moore, Stanford, 167
Morley, Edward, 67, 68
motion
    Einstein, 162–163
    Newton, Isaac, 17–18, 67
M-theory, 9, 40, 251, 263–266
    fiction, 296
    Maldacena, 276
    Sundrum, 280
    teaching of, 293
    testing of, 11–12, 274–275, 286, 292
muons, 177, 248
Mussolini, Benito, 135
Mussorgsky, Modest Petrovich, 27
mysticism, higher dimensions, 44–46

Nahm, Werner, 243, 248
Nambu, Yoichiro, 232
Napoleon (emperor of France), 30
Nappi, Chiara, 238–239

National Aeronautics and Space Administration (NASA), 272
natural forces
 described, 2–4
 interactions of, 61–62
 Kaluza-Klein theory, 4–9
Nazi Germany, 28, 153–155, 157, 165, 166, 168, 178, 183, 184–188, 193, 196–197, 215, 216, 229. *See also* World War II
Nee'man, Yuval, 223, 224
Nemorarius, Jordanus, 228
Neumann, John von, 191
neutralinos, 286
neutrinos, 248, 272
neutron, 176
Neveu, André, 234, 236
Newcomb, Simon, 49, 53, 59–60, 82
Newmann, Ted, 214
Newton, Isaac, 2, 17–18, 20, 38, 39, 54, 62, 67, 68, 69, 142, 287
Nicolai, Hermann, 249
Nightingale, Florence, 41
Nilsson, Bengt E. W., 250
Nishina, Yoshio, 115, 137
Noether, Emmy, 184
nonperturbative methods, 260–261
Nordström, Gunnar, 7, 86–87, 90–93, 105, 129
nuclear physics, 175–178, 183. *See also* atomic bomb

occult. *See* supernatural forces
octahedron, 13
Oersted, Hans Christian, 62
Olive, David, 263
Oppenheimer, Robert, 135, 208, 210, 213
optics. *See also* light
 electromagnetism, 64–65
 research in, 66
Osmond, Humphry, 202
Ostwald, Wilhlem, 118
Ouspensky, P. D. (Peter), 46
overdetermination, 173
Ovrut, Burt, 278, 279, 284

Pais, Abraham, 8
parallel postulate, 14, 31
parallel universe, 271–272
paranormal experience, fourth dimension, 46. *See also* supernatural forces
parapsychology, 200–202. *See also* supernatural forces
particle theory, 222–230. *See also* specific theories and particles
Pascal, Blaise, 228
Pati, Jogesh, 227
Pauli, Wolfgang, 6, 7, 115, 121–122, 125, 128, 130, 135, 137, 138, 149, 151–152, 165, 166, 174, 176, 191–196, 197, 198, 199, 200, 201, 208, 215, 217, 218, 238, 291
Pauli principle, 130, 232
perception
 hyperspace, 50–53
 parapsychology, 200–202
Perrin, Jean-Baptiste, 118
perturbative methods, 260
photons, 222, 223, 265

Picasso, Pablo, 81
Pickford, Mary, 139
Pickover, Clifford, 296
pion, 177
Planck, Max, 204
Planck scale, 243, 245, 246
Planck's constant, 6, 126
plasma physics, 228
Plato, 12–13, 72, 187
Platonic solids, 13–14
Playfair's axiom, 14
Podolsky, Boris, 162
Poincaré, Henri, 72–73, 83
Poland, 180
Polchinski, Joseph, 261, 264
Polish Intellectual Cooperation Committee, 180
Pope, Christopher N., 250
Pound, Robert, 267
Price, H. H., 202
Price, John, 288
Princeton Institute for Advanced Studies, 152–153, 155–156, 159, 166, 184, 188–189, 193–196, 235
probability, 172
projective relativity, 159, 160, 193, 198, 238
psychedelic experiments, 202
psychics, fourth dimension, 45–46. *See also* supernatural forces
psychoanalysis, 192, 200
Ptolemy, 13, 19
Pythagoreans, 13
Pythagorean theorem, 35–36, 77, 89

quantum chromodynamics (QCD), 3, 224, 234, 291–292
quantum electrodynamics (QED), 2, 208–210, 224
quantum theory
 Bohr, 145–146
 Einstein, 142–143, 144–145, 146–147, 151, 162, 171, 174, 256
 Feynman, 207–208
 general theory of relativity and, 211–212, 258
 geometry, 258–259
 gravity, 239–240
 Heisenberg, 132–133
 Jordan, 136
 Kaluza-Klein theory, 6–7
 Klein, 115, 120–121, 123, 128, 131, 133, 136
 Nazi Germany, 185
 nuclear physics, 175–178
 Pauli, 193
 Schrödinger, 132–133
 strong interaction, 2–3
 Wheeler-DeWitt equation, 220
quarks, 222, 224, 232, 248
quarternion theory, 39, 63

Rabi, I. L., 177
radioactive decay, 2, 176
radio waves, 66
Randall, Dana, 279
Randall, Lisa, 268, 274, 279–281, 283, 285

Raymond, Pierre, 234
Reichenbach, Hans, 200
relativity theory. *See* general theory of relativity; special theory of relativity
religion, hyperspace, 44
Renaissance art, 15–16, 17
Reynolds, Cecil, 141–142, 144, 150
Rhine, Howard Banks, 201
Riemann, Georg Friedrich Bernhard, 25, 32–37, 38, 39, 40, 42, 43, 86, 91, 104, 203, 247, 292
Riesenfeld, Ernst, 118
Robbin, Tony, 294, 295
rock climbing, 278–279
Rohm, Ryan, 255
Rome (ancient), 15
Roosevelt, Franklin D., 178
Rose, Peter, 294–295
Rosen, Nathan, 162, 217, 296
Rosenfeld, Léon, 180, 210, 257
Rovelli, Carlo, 258–259
Rowling, J. K., 49
Rubakov, Valery, 282
Rubin, Mark A., 244, 245
Rucker, Rudy, 54, 295
Rust, Bernhard, 185

Salam, Abdus, 4, 116, 183, 206, 220, 226, 227, 228–229, 230, 233, 238, 246, 247, 248, 249, 254
scalar, 90
scalar-tensor theory (Jordan-Brans-Dicke model), 198–199, 206
Scherk, Joël, 230, 235, 236, 239, 242, 243, 244, 248, 249
Schild, Alfred, 214
schizophrenia, 202
Schlegel, Victor, 48, 187
Schofield, A. T., 44
Schrödinger, Erwin, 115, 116, 128, 131, 132, 133, 144–145, 146
Schrödinger equation, 115, 125, 128, 145
Schucking, Engelbert, 199, 210, 219, 258
Schwarz, John, 233–234, 235, 238, 239, 243, 244, 250–251, 253–254
Schwinger, Julian, 182, 208–210, 212–213, 224, 239
science fiction, 203–204, 294–295
sculpture, three-dimensional art, 16
selectrons, 286
Sen, Amitaba, 258
Sen, Ashoke, 263
Serling, Rod, 202
Seurat, Georges, 240
Sezgin, Ergin, 261
shadow, light and, three-dimensional art, 16
Shakespeare, William, 54, 103
Shaposhnikov, Misha, 282
shrinkage, 246
Siegel, Warren, 250
singularities, 162, 172
Six-Day War, 229
Slade, Henry, 44–45, 65, 142
Smolin, Lee, 258, 259
Smythies, John R., 201–202
Society of Psychical Research, 46, 142, 201

Solvay Institute conference, 145–146
Sommerfeld, Arnold, 80
space
    gravity, 2
    non-Euclidean geometry, 40, 43
    special theory of relativity, 70–71
space-time
    Einstein, 151, 216
    Kaluza, 109, 110
    Minkowski, 73–78, 84, 87, 88
space travel, 296–297
special theory of relativity
    gravity, 85
    light, 69
    Minkowski, 71–80
    space, 70–71
    time, 70
Spinoza, Baruch, 138
Spiropulu, Maria, 288
spontaneous symmetry breaking, 225–228
squarks, 286
Stachel, John, 8, 79, 94, 158
Stanford Linear Accelerator Center (SLAC), 269
Stark, Johannes, 196, 197
Stein, William H., 167
Steinhardt, Paul, 283, 284, 285
Stelle, Kellogg, 261
Stewart, Balfour, 65
Stewart, Ian, 57, 295–296
Strauss, Charles, 293–294
stress-energy tensor, 88, 111, 150, 198
Stringham, W. Irving, 47–48, 49, 82
string theory, 202, 262–263, 268. *See also* brane models; superstring theory
    cosmology, 285–286
    gravity, 235
    hadronic, 232–234
    Kaluza-Klein theory, 276
    loop theory compared, 259
    Maldacena, 276–278
    M-theory, 263
    superstring theory, 9, 217, 234, 244, 250–257
Strominger, Andrew, 255, 274
strong forces
    natural forces, 2–4
    subatomic physics, 222–224
    supersymmetry, 232, 233
subatomic physics. *See also* specific theories and particles
    Kaluza-Klein theory, 206
    strong forces, 222–224
    supersymmetry, 231–233
    weak forces, 224–228
Sundrum, Raman, 279, 280–281, 283
supergravity, 9, 239–244, 246, 249–250, 251
supermembranes, 261
supernatural forces, 9, 26–27, 44–46, 142, 151, 200–202
superspace, 232
superstring theory, 9, 217, 234, 244, 250–257, 262, 285. *See also* brane models; string theory
supersymmetry, 231–235, 239, 240, 249, 261
Susskind, Leonard, 277

Sylvester, James J., 40–43, 46–47, 49, 55, 57, 72, 82
Sylvester, Sylvester J., 41
symmetry breaking, spontaneous, 225–228

tachyonic cuts, 233
Tait, Peter, 65
technology, higher dimensions, 296–297
telepathy, 200–202. *See also* supernatural forces
temperature, 76
ten-dimensional superstring theory, 256
tensors, 76–77, 87–88, 90, 198
tesseract, 50–51, 57
tetrahedron, 13
theory of invariants, 41–42
Theosophy movement, 45–46, 82, 292
Thiry, Yves, 199, 206
Thomson, J. J., 72
't Hooft, Gerardus, 227, 234, 277
Thoreau, Henry David, 63
Thorne, Kip, 296
three-dimensional art, 15–16
three-dimensional geometry, Greece (ancient), 13–14. *See also* Euclid; geometry
time
    fourth dimension, 18–19, 57–60, 71, 73
    gravity, 2
    Newtonian physics, 69
    special theory of relativity, 69–70
The Time Machine (Wells), 58–59
Tolkien, J. R. R., 267
Tolman, Richard, 135, 138, 150, 155, 157, 189
Tomonaga, Itiro, 208, 210
Tonnelat, Marie Antoinette, 218
Townsend, Paul, 261
transrealism, 295
Tschirnhaus transformations, 104
Turok, Neil, 284
The Twilight Zone (television series), 202
two-dimensional art, 15

Uhlenbeck, George, 130–131
uncertainty principle, 133, 143, 145, 208, 211, 288
unidentified flying objects (UFOs), 201
unified theory
    Einstein, 111, 147–148, 149, 159, 172, 216, 282
    electromagnetism, 66
    Kaluza, 108
    Nordström, 91
    Pauli, 192–193
    quest for, 60, 62
    Riemann, 33, 38
    subatomic physics, 222–224
    teaching of, 292–296
    technology, 296–297
    Weyl, 96–99, 128
    Yang-Mills gauge theory, 221
universe. *See also* cosmology
    astronomy, 17
    Newtonian physics, 17–18

van Dongen, Jeroen, 174
van Nieuwenhuizen, Peter, 227, 239, 250
Veblen, Oswald, 159, 160, 161, 193, 268
Veneziano, Gabriele, 232
Verne, Jules, 49
virtual photons, 208

Walpurgis Night, 26–27
warp factor, 280
waves
    Broglie, Louis de, 123–124, 126
    electromagnetism, 64–65
weak forces
    natural forces, 2, 4
    subatomic physics, 224–228
Weber, Max, 81
Weber, Wilhelm, 33, 34, 62, 72
Weinberg, Steven, 4, 189, 226, 227, 230, 231, 246, 247, 248, 251
Weisskopf, Victor, 188
Wells, H. G., 18, 49–50, 58–59, 71, 82
Wentzel, Gregor, 166
Wess, Julian, 234, 239
Wesson, Paul, 282
West, Peter, 249
Weyl, Hermann, 96–99, 102, 105, 108, 110, 111, 112, 128, 148, 149, 166, 181, 184, 185, 193, 201, 217
Wheeler, Janette, 169
Wheeler, John, 89, 169, 177, 196, 207, 214, 219, 220, 259, 277, 296
Wheeler-DeWitt equation, 220, 257, 258
Whitman, Walt, 63
Wigner, Eugene, 180
Wilde, Oscar, 46
Wilkinson Microwave Anisotropy Probe (WMAP), 272, 283
Wiltshire, David, 282
witchcraft, 26–27, 44. *See also* supernatural forces
Witten, Edward, 236–239, 247–249, 251, 254, 255, 262–263, 264, 270
Witten, Louis, 237
World War I, 107, 118, 184
World War II, 178, 183, 188–191, 197. *See also* Nazi Germany
wormholes, 296–297
Wünsch, Daniela, 102, 185

Yang, Chen Ning (Frank), 182, 221, 223, 228, 234, 239, 248
Yang-Mills gauge theory, 221, 223–225, 227, 240
Yau, Shing-Tung, 256
Yukawa, Hideki, 116, 177, 180, 181, 223

Zeeman effect, 130, 131
Ziegenbein, Paul, 185
Zöllner, Johann, 44–45, 46, 83, 142, 201, 292
Zumino, Bruno, 234, 239, 241
Zweig, Georg, 222, 224